Salvage from the Sea

By the same author:

Long Distance Swimming
First Strokes in Swimming
Modern Long Distance Swimming
Brown's Pocket Book for Seamen

To Charles and Vega

May your own careers give you as much real pleasure
and immense job-satisfaction as mine has to me

Salvage from the Sea

Commander Gerald Forsberg, OBE, FNI, RN
Master Mariner

GLASGOW
BROWN, SON & FERGUSON, LTD.
4-10 DARNLEY STREET

First Edition - 1977
Reprinted - 1992

ISBN 0 85174 606 3

Printed and Made in Great Britain

Contents

Illustrations

Foreword

by Captain Ian M. Clegg, CVO, FNI, RN, Master Mariner

Most mariners would agree that abandoned wrecks are sad and thought-provoking sights. They are dismal monuments to human failures and to lost battles with the elements. This book describes clearly and simply the work of those dedicated professionals engaged in 'salvage from the sea' and how they play a part in making such sights comparatively rare.

The author is well fitted for his self-appointed labour of love. He is a first-class practical seaman with a long and varied experience and has been in the salvage 'business' for twenty-five years or so.

Perhaps his first introduction to this branch of seamanship was the refloating of a 39-ton ketch which ran ashore in the Solent when he, aged fifteen, was a member of the crew. Incidentally, immediately after being praised by the Captain Superintendent of the Training Ship to which the ketch was attached for his efforts in saving her, the author was awarded twelve strokes of the cane for an infringement of the Training Ship's rules during the salvage operation. Some years later he played an important role in the salvage of the Comet airliner which crashed in the sea south of Elba in 1954 for which he was awarded the OBE – and no strokes of the cane!

The book is small and obviously cannot cover such a vast and complex subject in detail. Nevertheless I believe it fulfils the aims wisely set out by the author in his preface. It gives an absorbing glimpse of the fascinating subject of salvage and, perhaps most important of all, it highlights the need for the ingenuity, tenacity and sustained persistence of salvors.

Preface

Disaster strikes quickly at a ship. A moment's error causes a collision. A careless action causes a major fire or an explosion. Fatigue, or human failure, can get the bottom of a ship ripped out in seconds. Typhoons and hurricanes can drive ships so far up a beach that they are high and dry when the weather improves. In wartime, torpedoes, bombs and mines cause instant marine casualties. As that great writer, Joseph Conrad, said: 'I have known the sea too long to believe in its respect for decency'. The air is no different from the sea as regards decency. Aircraft ranging in size from small helicopters to enormous airliners occasionally plunge from sky to sea in moments. They often need recovering to establish the cause of accident and to prevent another similar accident.

To amplify what I have written above, take a look at the chapter headings. You will see that there are many types of marine casualty. And each type divides into sub-types. In essence, every salvage case is unique; there can never be 'standard' procedures. The only way to proceed is to view every case with knowledge of general physical principles and with the widest possible experience as a seaman. Then to use that knowledge and experience to make an operational plan – but subsequently to use common-sense to amend/scrap/use that plan hour by hour if necessary.

In the *Journal of the Royal Institute of Navigation* (vol.28 no.2) Professor P.K. M'Pherson said that in this maritime nation of ours there was an almost complete absence of sea-topics from educational syllabuses. How extraordinary that is in an island nation whose history has shown how dependent it is on the sea for its food, commerce and defence.

Professor M'Pherson was not referring to specialist courses in naval architecture or ocean science but the absence from education at large in schools and universities of courses connected with the oceans and British maritime interests as an essential component of any general educational programme for the future citizens of our islands. The Professor is undoubtedly and entirely right. The aims of this book are therefore four-fold:

(a) to be read in schools. True the style is somewhat racy but good instructive books do not *absolutely* necessarily have to be stuffy in presentation;

(b) to give a pleasurable insight into the subject for the general reader who has an enquiring mind and likes to enlarge his knowledge beyond the restricted confines of his own immediate tasks;

(c) to 'put in the picture' a vast number of worldwide seafarers, engaged in absorbing seamanlike pursuits of their own, as to how one other section of the profession earns its living and achieves immense job-satisfaction;

(d) to assist salvage officers, even if ever so slightly by mention of perhaps just one unknown fact; by acquainting them of just one useful book to read; by giving just one crumb of consolation that, whatever setbacks they are currently experiencing, someone somewhere has endured them before.

The way I hope to fulfil those aims is to outline some of the problems, to describe briefly the techniques in plain understandable language and to spin a few salty yarns. I hope particularly that it will all be appreciated by those who like information served up without pomposity, without exaggeration, without ballyhoo and, above all, without loud trumpets sounding on behalf of the author. The whole subject of salvage and recovery is fascinating beyond all; it is laced with superb seamanship, cool nerves and, most importantly, a great amount of solid horse-sense. I love the subject dearly myself and hope to pass on some of that feeling to you.

You will perhaps notice that I have omitted putting in a special separate chapter on fires in ships. It is an entirely purposeful omission because the general end-product of fire is a towage, righting or refloating job which is described in various other chapters and does not differ in execution because the reason for sinking, capsizing or immobilisation is by fire rather than from some other cause. The fighting of fires at sea is the responsibility of the ship's officers who have been trained in firefighting and know their own ship's equipment better than any stranger could possibly do. Fires on ships in port (and in United Kingdom coastal waters) are the responsibility of the nearest regional Fire Authority. The best aid that any salvage vessel can give is to batten down and cut off the supply of oxygen to the fire, to cool the affected parts by constant hosing, to inject foam into affected compartments or, sometimes, to judiciously let the fire burn itself out before commencing other salvage work. Tugs can and do hold burning ships in such a position relative to the wind to minimise the spread of fire and to ease the work of the firefighting teams.

Here then is my personal presentation of *Salvage from the Sea.* I know well, simply because the book is aimed at an intelligent readership, that occasionally you will be frustrated by the inevitable lack of depth due to the book's small size. For this particular reason, I have given an extensive list of references at the end of the book and I feel sure that they will lead you on to all the depth you require.

As it is impossible to avoid odd bits of technical and professional jargon, I have tried to explain them as they first appear in the text. To have attempted a complete nautical dictionary would have been quite impracticable and anyway so many people mess about in boats nowadays that they are well aware of such terms as bow and stern, fore-and-aft, thwartships and many such others. For good measure, however, refer to the *Dictionary of Nautical Words and Terms* (see chapter 14).

Finally, with the very gradual transition of this conservative country (and its even more conservative seamen) but by no means by all important maritime nations, to the metric system, it has been difficult to attain consistency in statements in lengths, tonnages, water depths, etc. In general when describing elderly cases or quoting from reports, I have used the Imperial measures in practice at the material time; when speaking of the present or future, I have used the Metric system. That is a typical British compromise but I could otherwise see no way of being reasonable in presentation.

Author's Prefatorial Note to Second Edition

I am happy – as inevitably as any writer in similar circumstances would be – to know that the first edition of this book has completely sold out. As it had the minimum of supportive advertising it has evidently been sold by word-of-mouth recommendation and that is, to me, the most pleasing aspect of all.

I now approach the task of preparing a second edition after an immense amount of thought on the best way of tackling it.

In essence I have decided to retain the bulk of the text as it was in the first edition on the grounds that a sell-out must indicate reader-satisfaction. In any case, it was inextricably bound in with my own personal story of salvage events and occurrences as I had known them (or become acquainted with them) in my then quarter-century of being connected with the salvage industry as such. After all, this is not a highly technical treatise but a virtually non- technical broad brush account for any intelligent reader to comprehend.

In any case, I feel very strongly that many of the elderly cases and anecdotes are well worthy of retention as sign-posts to the state of the salvage art at various dates. If I delete such accounts, they may disappear for ever — a severe and unnecessary loss to future students, historians and researchers.

Some forecasts made in the first edition have been

overtaken by events. Some minor gremlin-type errors which crept in need to be expunged. Some new instrumentation and operational tools have evolved and deserve a note. All these items will be noted and dealt with.

But, in the main, the book will remain with the same format and content as has pleased the readers so much for so long. I hope you enjoy reading it as much as I enjoy putting it together.

Acknowledgments

I am most grateful to the United States Naval Institute for permission to use a small but valuable amount of copyright material from its *Proceedings*. This debt is added to by the considerable amount of general sea knowledge picked up from its pages over the years and which it is impossible to acknowledge. My specific thanks go to the authors of the following articles: *An Unusual Case of Ship Salvage* by Commander Thomas N. Blockwick, USN; *The Valor of Inexperience* by Captain Harvey Haislip, USN; *Contact 261* by Captain Lewis B. Melson, USN; *USS Thresher (SSN-593)* by Vice-Admiral E.W. Grenfell, USN. Nearer home, the diary of the late Doctor F.W. Hogarth was of value in putting on record an historical refloating case of the early twentieth century.

In the matter of illustrations, I cannot sufficiently thank those who delved so unselfishly and readily into their collections to help me – The Mersey Docks and Harbour Company, Allan C. Crothall, Esq., (lately of Risdon Beazley/Ulrich Harms), the United Towing Company, Vickers Oceanics, Mrs. Elizabeth MacNeice, Tony and Marion Morrison, Skyfotos Ltd, Stewart Bale Ltd, *Evening Gazette*, Middlesborough, the *Argus*, Capetown.

High above everything else, I have appreciated working with the small corps of Admiralty Salvage Officers. Their professional expertise and personal friendship have given me, both afloat and ashore, the happiest and most satisfying quarter-century of my maritime career.

Finally, as with my other books, this one would never have seen the light of day without the patient encouragement and typing skills of my wife. Frightfully disrespectful but very useful criticism has also been made by my daughter.

Second Edition. Further Acknowledgments.

For continuing advice on the state-of-the-art concerning underwater tools, electronic navigational devices, sonar advances and so on, thanks to my son, Charles, sometime of the Royal Naval Hydrographic Service and currently serving with a world-class firm in the underwater industry.

But, above all, boundless gratitude to Captain Bill Searle, US Navy (Rtd) formerly Supervisor of Salvage with that Navy, later head of a consortium dealing with almost every aspect of ocean engineering and, finally indefatigable collector, monitor, recorder and custodian of vast amounts of knowledge on underwater engineering activities. He has answered my every call for assistance – sometimes instantly from his own experienced memory and sometimes after consulting top experts on my behalf. But I never drew a blank.

Gerald Forsberg

Chapter 1

What salvage is all about

Let's kick off with two essential definitions. I apologise for being semi-legalisitic at the very outset but otherwise the salvage purist will be snapping at our heels for not separating 'true' salvage from other forms of recovery work. There are almost as many legal definitions of salvage as there are seagulls on the white cliffs of Dover. But the ones I give here contain the basic essentials for non-specialist readers. Legal readers please go direct to Kennedy's *Law of Civil Salvage* which is, both in British and United States courts, regarded as the most authoritative volume in the business.

A Salvage Service is a voluntary service given to save from peril a ship, cargo or other property concerned in a marine adventure.

A Salvage Award is payable for a salvage service out of the value of the object or objects saved.

Having set that bit of record straight, we can leave discussion of legal and commercial aspects to a later chapter of their own. Until then, we can stick to the operational aspects of salvage and for the sake of completeness also deal with those other maritime recovery operations which do not legally rank as salvage. That is the point at which purists snap most viciously. But it is entirely logical to deal with salvage and recovery together; the two sets of operations use similar ships, personnel, equipment and must take on the same harsh elemental enemies – tides, weather, sea conditions and climate.

First, then, one positively fundamental historical statement. In former days, every shipmaster had to be his own salvor. When sailing vessels and horses were the most rapid form of transport – and semaphore the fastest communication system – there was absolutely no time to await specialist advice. Prompt action was imperative by the man on the spot; this was not only necessary to combat tide and weather conditions but also to thwart the hostile designs of local inhabitants, even in civilised countries, who very much favoured wreck and abandonment on their front doorsteps. Indeed it, has been said that if all former wrecks could be restored to service, and could demand back all their stores and materials, half the houses in Cornwall would fall flat to the ground. I dare say it goes equally for many other places.

On the subject of communications, it is interesting to note the Canute-like attitude of the British Government to progress and improvements of facilities. In 1823, it instructed its obedient servants in these terms: 'Telegraphs of any kind are wholly unnecessary and none other than the semaphores presently in use should ever be used'. It is good to record that times and attitudes have changed; present-day governments are *never* stick-in-the-mud and reactionary. And if you believe that, you will believe anything!

In those former times, a shipmaster was a seaman high above all else. (Every finger a marline-spike; every hair a rope yarn!) Today a shipmaster is heavily burdened with administrative and managerial responsibilities which inevitably distract him from the purely seamanlike aspects of his duties. Besides, the size of the task has changed. For instance, in 1580, the *Golden Hind* was refloated after stranding by means of jettisoning eight pieces of ordnance and three tons of cloves. But now, towards the beginning of the twenty-first century, the task of refloating a giant tanker or ore-carrier is somewhat less straightforward; it requires a tremendous amount of specialist skill backed by specialist vessels, machinery and equipment.

The main thing which, however, sounded the death knell of do-it-yourself salvage was the ever increasing speed of transport. First railways and steamships; then aircraft, helicopters, hovercraft and hydrofoils. Today there is practically nowhere in the world a salvage officer cannot reach in a couple of days – although I do remember a Salvage Association surveyor coming to the end of modern transport and having to walk the last twenty miles in Ethiopia. Further by modern communications systems, legal and operational advice can be on its way within minutes.

Mind you, the end product of modern communications is not always uniformly useful. I recall a British merchant vessel which had the misfortune to run hard aground on a small Pacific island. And despite every strategy in the book, she was quite unable to refloat herself. We were somewhat surprised to see a telegram winging its way to the ship from her owners. 'Have you', the telegram enquired, 'tried putting your engines full speed astern?' I think the Master was a bit surprised too. As for the salvage officer, he turned into a savage officer forthwith.

The first recorded ship casualty was, of course, the *Ark*. The Atlantic Mutual Insurance Co. of New York has tremendous pride in the completeness of its casualty records; asked by a correspondent for information on the *Ark* it reported back: 'Built 2448 BC. Gopher wood, coated with pitch within and without. Length, 300 cubits; width 50 cubits; height 30 m cubits. Three decks. Cattle carrier. Owner: Noah and Sons. Last reported stranded Mount Ararat'.

I do not intend a long-winded chronology of marine salvage from the *Ark* onwards but a few selected examples will serve well to set the scene for understanding modern operations.

Apart from that early unsuccessful case, ship-salvage has been practised since time immemorial and so has the associated pursuit of diving. Alexander the Great (356-323 BC) became the forerunner of all modern submersible operators when he had a glass bathyscape made; a manuscript at the British Museum in London shows him at work in it – comfortably lying down but uncomfortably wearing his crown.

Herodotus, in 460 BC, speaks of a diver recovering treasure from wrecked vessels. The Rhodians, when dominating the Eastern Mediterranean *circa* 300-200 BC, had a scale of rewards for divers according to depth. Livy (59 BC-AD 17), writing much later, states that divers recovering items from 8 cubits (4 metres) received one third of the value; recovery from 16 cubits entitled the salvor to half value. That same aquatic Alexander the Great employed divers to destroy the boom defences of the port of Tyre in 333 BC.

In above-water spheres too, we know of numerous well-known feats of salvage from the past. In the field of ocean towage, sailing vessel frequently towed sailing vessel for long distances – an almost unbelievably feat of high seamanship for modern salvors to contemplate. Sail-tow-sail operations are recorded in several old logbooks of which Anson's 1741 *Journal* is one. Ships a-plenty damaged by fire, explosion, weather or action were beached, careened, repaired and refloated for services by ships' carpenters and their mates.

The times when the appropriate material for ship-repairs grew on trees which nobody owned really were 'the good old days'. Nowadays flying out and fitting, say, a 70-ton propeller for a ship presents different problems. Nor is ocean towing the same affair of impromptu inspired seamanship as in sailing ship days; a tanker of half a million tons deadweight requires more than the goodwill of a passing mariner to urge her in the right direction.

Getting more to recent times, maritime casualties in war are inevitable. The enemy is trying to inflict maximum damage – by guns, mines, torpedoes, frogmen, bombs, guided missiles, depth charges, and anything else he can dream up – and it is evident that a proportion of the attacks must get through. The 100 per cent perfect defence has not been invented yet. In this connection it is well worth putting on record the appreciation of the greatest Briton of modern times. In February 1941, Prime Minister Winston Churchill wrote:

I learn that the salvage organisation has recently made as big a contribution to the maintenance of our shipping capacity as new construction, about 370,000 gross tons having been salved in the last five months of 1940 as against 346,000 tons built; while the number of ships dealt with by the salvage operation has increased very rapidly from ten in August to about thirty now. They are to be congratulated on this . . .

For me looking back into the past is a fascinating and absorbing exercise very worth a whole book in itself. But long and sorrowful experience has made me aware that everyone does not wholeheartedly share that enthusiasm: it has made me aware that the appearance of glazed eyes and the symptoms of switched-off ears is the cue for me to change subjects without delay – or faster if possible. So, especially as this book has to cover a wide range of subjects in short space, we will definitely not tarry too long in the world of long ago. As each chapter starts, I will merely take a quick run through the beginnings and then get on with the present – with one proviso. Good cases are good cases irrespective of their age. It is by studying the lessons learnt from earlier cases that the standard of salvage practice will grow better and better in the future. We hope! Any way history only stopped a micro-second ago and someone somewhere is making it at this precise moment. For the confirmed Henry Ford ('history is bunk') type readers, I will be careful to refer to 'experience' or 'long practice'. The sweet smell of a rose – so they say – remains the same even if referred to as a cabbage!

In the piping days of peace, casualties are not inevitable. But they somehow happen all the time. In many years of reading the daily newspaper *Lloyd's List*, I cannot recollect any single day that the casualty columns were blank. In similar fashion the Institute of London Underwriters produces a monthly and annual return of casualties and neither of them ever lacks for ample material. One typical annual return, taken at random, records the following total losses: weather damage, 26; founderings and abandonments, 23; strandings, 48; collisions, 19; contact damage, 8; fires and explosions, 54; missing, 4; damage to machinery, 3; other causes, 10. Total losses, are of course, those extreme cases where salvage has either been impracticable or has failed. The list of vessels where salvage has been practicable and has not failed runs into many hundreds.

You may wonder how so many casualties can possibly occur in a well-regulated modern world? Sometimes they are reprehensibly caused by grave dereliction of duty such as an officer having left the navigating bridge without proper relief. Now and again, ships genuinely run on to an uncharted rock or shoal – although normally this reason is received with the greatest incredulity in the profession. Masters and officers often have to spend inordinately long periods on duty and mistakes due to fatigue may occur – or they can suddenly be taken ill. Unfortunate ships can stumble ill luck on a hitherto unknown scientific fact; for instance that washing out oil tanks with high pressure water jets induces a sufficiently dangerous electrical charge into the ships structure to cause a disastrous explosion.

The list continues in a variety of ways ranging from act of man to act of God. Sometimes a collision is caused by some foolish 'rogue' ship going the wrong way through a shipping separation zone in order to save a few minutes on time of arrival in port. (At the time of writing the first edition it was estimated that one ship in every seventy was going through the Strait of Dover against the traffic flow. Even today, after long familiarity and use, there is still the occasional 'rogue' ship to cause a flutter in the Coastguard dovecotes). Many casualties are caused by the sheer fury and violence of weather – not only at sea but, say, Hong Kong harbour during passage of a typhoon. Sometimes a ship breaks in two because a structural weakness has been waiting the right conditions to develop into a structural failure. In the act-of-man category, comes the mad fool who smokes in a 'no smoking' area of the ship. Uniquely there is a ghastly unpredictable misadventure such as when the *Lake Illawarra* struck the Tasman Bridge at Hobart, flung crossing cars into the river and brought down a huge concrete span on top of herself. In case you are making a mental resolve never to travel by sea again, it must be said that there are some 200,000 ships at sea. The number of casualties looks far worse than the percentages!

In addition to ship accidents, aircraft add their quota to the marine salvage burden. Great airliners drop from sky to sea in a matter of moments. Prototype military aircraft disappear from the radar scan. Helicopters make forced landings in the sea. Unless the wreckage of these aircraft is recovered for inspection and testing there is always a strong chance of another accident happening to a similar aircraft. Ignorance of the cause of an accident is bad for the morale of manufacturers and operators; its worse still for the morale of aircrew and passengers. Thus, after almost any air disaster or important accident over the sea, the call goes out to some salvage organisation. Even if the aircraft is in a thousand pieces (which it often is) the accident investigators will piece them together like a jigsaw and produce an accident diagnosis.

To all that catalogue of tasks for salvors, must be added all the more routine, minor-league, work. Human nature being what it is, somebody in some place is regularly losing something underwater. It may be a toppled crane, a runaway locomotive, a harbour lighter or some sort of valuable missile. The last task may sometimes escalate into a major-league operation. On a more mundane plane, there is a fairly regular job picking up anchors and cables which, for some reason or other, ships have had to leave behind on the seabed. Scientists want 'black boxes' lowered to the sea bottom; others of their colleagues want other black boxes raised to the sea surface.

Naturally there is a strong school of thought that to have no accidents would be better than having any number of splendidly successful salvage cases. On the premise that prevention is better than cure, government departments of many nations labour mightily at improving the safety of their ships. Rigorous judical enquiries are held by lawyers assisted by nautical assessors and appropriate recommendations are made for the future. In the international field, the International Maritime Organisation (IMO) which is a specialised agency of the United Nations labours equally mightily. But theirs is an uphill task. International agreements take much time; full ratification takes more. And in the meantime, new types of risk arise, more complicated vessels and maritime structures appear on the scene. Above all, a new generation of accident-prone men are appointed to positions of responsibility.

The safety organisations have a gloomy outlook. The best they can hope for is to maintain the accident rate much as it is; complete victory and a casualty-free maritime world can never possibly be theirs. It is on rather a surprising note that it must be reported that the profit motive plays a considerable part in the battle for additional safety measures. The great and influential body of international marine insurance is a powerful ally. For underwriters are determined not to be out of pocket after settling claims and shipping firms are equally determined not to have to pay large extra premiums to insure bad risks. Thus there is a joint support for all possible safety measures.

Despite all that thought, work and action on prevention aspects, the hard fact remains that demand for maritime salvage and recovery work seems never ending. The main sadness for salvors is not shortage of work but that mishaps never happen in an even flow. Either their ships and employees are working at incredible pressure for twenty-four hours a day, seven days a week or they could be uneconomically hanging about waiting for the next accident to happen.

For that strong economic stop/start reason entirely, there

are practically no single-purpose salvage firm pure and simple. Almost every firm in the business likes to have a long-term 'bread and butter' job like contract towing, contract mooring work or attendance on offshore oil rigs. They can then juggle with their fleet of vessels and divert one or more to obtain their salvage 'honey' when the opportunity occurs. There are a very few salvage vessels and tugs around the world on 'salvage station' but they are almost always units of a large fleet which has its other vessels on programmed tasks. For a firm to attempt existence in ordinary peacetime conditions on salvage awards alone is a mighty tough option and a somewhat unnecessary one when several types of maritime work can be done so well in parallel. In summary, if ever there is a job for life, or a job for eternity, it is in the world of salvage, recovery and allied work.

There is always something new coming along. I have no doubt that one day the Lord High Panjandrum of Ruritania will demand the rescue of a ditched space-capsule from the bottom of the Challenger Deep in the Marianas Trench – say from 11,000 metres. And I have equally no doubt that one salvor – after many other salvors have declared it quite impossible – will get it back in due course.

Chapter 2

The Salvage Officer. Salvage personnel. Some problems

In every enterprise, the personnel factor plays a part of paramount importance; in salvage and allied operations, this is particularly the case. Shipmasters, officers and divers regularly carry out duties in which human lives are at stake. Good petty officers, able seamen and engine-room ratings are also essential links in the chain of efficiency and safety. Carelessness by a boatswain (or by the chargeman of a shore salvage party) could, as just one instance, easily part a heavily stressed wire rope and take an arm, leg or head off an unfortunate colleague. And, since perfection is not given to man, it has happened. We have learned our lessons the hard way.

Luckily, although working and living conditions are pretty unattractive, there always seems to be a central hard core of devoted and well-trained salvage men available. Those who persevere in the salvage way of life are usually physically strong men imbued with much horse-sense and incredible patience. They have learned that weather and tides are almost invariably the complete arbiters and that attempts to beat them by hurry or by force usually end in disaster. And disaster may be the only possible word.

The ability and temperament of salvage officers is,

however, the greatest single factor to attainment of success. Occasionally it is the sole factor. In addition to the qualities just mentioned for other personnel, a first-class salvage officer needs much theoretical knowledge and much practical experience. These usually accrue as he progresses with time from second or third assistant to established salvage officers on big jobs, moving on to small jobs in charge (with a judicious but unostentatious bit of supervision from his peers in the background) and finally to tackling bigger cases on his own.

In the Navy of Nelson's time, the officers thinking of their own slim promotion chances used to drink an after-dinner toast: 'Here's to a bloody war and a sickly season'. Promotion is something like that in the salvage world except that the offshore oil and gas industry is temporarily offering professionally attractive employment. I say temporarily but it does depend on exactly how one defines that word! The enormous upsurge in offshore exploration and construction work must inevitably reach and pass its peak; then the routine production platforms will require only routine maintenance and support. But finally, at the end of their useful lives, every offshore structure will need to be dismantled and removed. At which point, albeit many years ahead, many of the redundant personnel will return whence they sprang; the alternative being to go off to newer, and even remoter and more hostile exploration areas. Anyone fancy a commission on the Antarctic continental shelf if all commercial activities have not been banned from there by that time.

The offshore industry is a wonderfully fascinating subject in its own right. It would be easy to digress but, having briefly shown its interface with the salvage industry I must be quite firm and leave it. It is, on the face of things, a competitor of the salvage industry for officers, seamen, divers and vessels. But wait. 'There are more ways' – so they say – 'of killing a cat than by choking it with butter'. And in pursuance of that wise philosophy, salvage and towing firms are now using the offshore industry to provide the essential bread-and-butter work for themselves. For years I wondered what 'industrial cross-fertilisation' meant. Now I know!

Back then to the young salvage officer; he can well feel frustrated by a long succession of humdrum, humble little jobs – like fishing vessels aground, a bit of heavy debris to clear from a harbour, a ship's propeller to burn off and carry ashore, a small wreck to cut up and so on. But one day a senior salvage officer will be sick or, as so often cussedly happens, two or three salvage cases will occur simultaneously at, say, the Falkland Isles, Famagusta and Frinton-on-Sea. Then the makee-learn officer gets a star role in five seconds flat. Its an ill wind…etc.

He also quickly learns that great perennial professional lesson of all time. Although he knew it all as an assistant, things look very different when he becomes entirely responsible for a fabulously expensive ship and the lives of all those working on the job. It is that timeless, ever-recurring, flash of light which comes to all seconds-in-command when they succeed to the hot seat. It is a particularly blinding light, however, to a young man in charge of his first salvage job in, perhaps, rather frightening conditions.

Marine salvage is not, and never can be, an exact science. The practical salvor is therefore no scientist but rather what the Immortal Bard (Scottish League, Premier Division) called 'a man o'pairts'. Nevertheless, of course, he has to depend on and understand many scientific principles in order to carry out the many varied tasks falling to his lot. First and foremost, though, he must be a good seaman; if he knows how to operate and keep afloat a seagoing ship, the chances are that he will have a pretty good idea of what is necessary to restore a sunken or stranded vessel to her rightful condition of buoyancy.

The need for theoretical knowledge and practical experience has already been mentioned. This must cover seamanship, maritime law, naval architecture, meteorology, tidal predictions and – to a certain limited degree – mechanical, civil and electrical engineering. If the salvage officer also knows something of hydrography and draughtsmanship that is a bonus. Good humour may not be essential but it is a great asset.

The Salvage Officer. Salvage personnel. Some problems

Salvage officers are rather like general medical practitioners. Both types are infinitely hard-working, painstaking bodies of men and they never know when the next emergency case is coming along. Work for ship salvors as well as 'people salvors' frequently comes in waves or epidemics. Rarely are salvors or medical men called in until every home-made remedy in the ship or house has failed and the patient is in a thoroughly bad way.

Salvage is frustratingly full of natural difficulties and far from devoid of man-made ones either. Good advice comes in profusion from spectators in the grandstand – be that edifice actually situated in the board-room or bar-room. Every boss and every professional colleague will probably know a 'better' way and not to hesitate to say so. The salvage officer also bears one extra great but inevitable burden; it is to suffer inventors gladly or as gladly as possible anyway. During every major incident (at the precise time when every salvor's whole being is intensely committed) countless ingenious and impracticable suggestions come leaping from active but landlubberly brains. In any well-organised outfit, the salvage officer should be able to fling these suggestions, with appropriate comments, to a backroom boy to deal with diplomatically.

As a backroom boy myself, I have dealt with very many of these inspired thinkers by letter, telephone or personal call. With each one of the inspired thinkers convinced that his personal pet idea is the only way to success, I reckon to be a master of 'the soft answer that turneth away wrath', but sometimes, after the fifth call from the same person on a hellishly busy day, urbanity has broken down. From experience, I know that to be the moment when the caller will threaten to complain about this unwarranted incivility to his 'great friend' the Member of Parliament for Loamshire, the Chairman of Lloyd's, the First Sea Lord or what-have-you. I once only rose to the dizzy heights of being threatened with being reported to the Prime Minister. And once, much more terrifyingly, it was to be to the Comptroller and Auditor-General! All those illustrious personages must be very understanding men for I was never on any occasion reduced in rank from Commander to Ordinary Seaman. Perhaps, though, I ought not to speak too soon?

From years ago, actually during the prolonged and difficult search for the submarine *Affray*, I treasure one outstanding suggestion. As ever, the telephone caller was convinced that he alone had hit on the correct answer. 'Line up all the divers in the United Kingdom on the south coast', he said, 'then let them link hands and march to France. Someone is bound to find the submarine on the way across'. The imagination boggles at the thousands of attendant diving boats (many without engines) trying to cross the Channel through the shipping lanes, in rough weather, athwart racing cross-tides and sometimes in darkness. To say nothing of the rather long walk for the divers.

Another gem from the depths of my elephantine memory, was when Port Said was being cleared of blockships in 1956-57. 'All that you need to do', advised this caller, 'is to install refrigerating machinery in the wrecks and freeze the water inside. Then, as ice floats, the wrecks will come to the surface and be ready for towing away'.

As a matter of fact, that idea is an example of a sound physical principle which could well have worked in a laboratory or even in a tank test with models. But on the actual job, with hundreds of thousands of tons of ironmongery on the harbour bed (and with no enormous amount of refrigerating machinery to be had) the concept is as useless as a busted flush. Some suggestions are not even laboratory worthy. Others proudly presented as brand, spanking new by their sponsors, were sometimes first sketched by the staff of Alexander the Great, Philo of Byzantium, Aristotle or Leonardo da Vinci.

Speaking, as I am ashamed to say, as a chap who has been a backroom boy longer than a frontroom boy, perhaps I thump the backroom table a bit. But the rear link (ashore in depot or office) has an extremely important part to play in furthering the salvage officer's endeavours. It can forecast requirements in equipment, stores and personnel and get them transported to the site. It can keep off the salvage officer's back the never-failing mass of enquirers asking: 'How long? How come? How much?'. The shore link can work out in comparative tranquility those difficult sums involving problems of structural strength, ship-stability, highest forecast tides, inherent dangers in the cargo, values, what amount of money to ask for as a bond, etc.

Above all, it is the shore link which rises to the occasion when the salvage officer demands 20,000 gallons of fire-fighting foam by tomorrow or sooner. Or when Diver Bloggs is urgently required to be flown to Micronesia but whose passport lapsed last week, who requires revaccination and a couple of other medical jabs and whose wife says: 'He is out and I do not know when he will be back'. One other person, ultimately beneficial but currently irritating to the man on the job, is the Finance Officer; his perpetual question is: 'Do you really need it or are you being emotive?'. It's the sort of question that keeps businesses profitable. There is no doubt that the shore organisation does pull its full weight in assisting the man on the job.

But let's return to the sea and the salvage crews. They are strange characters and considering their life, they could hardly be otherwise. They spend perhaps ten months of every year cooped up in a poky little ship with, if lucky, a few nights and weekends ashore. In many other types of ships, ten months could mean a world circuit with a run ashore at Capetown, Rio, Montreal or some other glamorous city. For the salvage crews, the routine is entirely different.

The salvage cruise-ticket is for dirty, hard labour on in-port tasks, uncomfortable solid grafting close in to some inhospitable coastline or even some long, slow, difficult tow across endless stretches of empty ocean. No Capetown, Rio, Montreal stuff for them – more likely to be Namsos, Christmas Island or Great Yarmouth. That's if they are lucky. If unlucky, they might take time out in some isolated uninhabited bay at the back of nowhere. Like that place the celebrated wartime new naval recruit wrote home about; 'Dear Mum, I cannot tell you where I am. I don't know where

I am. But where I am there is miles and miles of b-gg-r all. Love, Ted'.

In small ships with thirty-odd officers and men aboard, one does not get Ritz conditions. There are minute cabins, communal mess-rooms, a small library, perhaps a small film library, a radio and a television. In any case they frequently turn the television off, radio off, ignore the other amenities and talk about salvage. Even hot and cold water may be rationed. Away from port, bread is dependent on weather being calm enough to bake; otherwise it is right back to bluenose, shellback, hard biscuit. In such a small community, living cheek by jowl for such long periods, shipmates' habits can begin to jar on the nerves of the susceptible.

One officer confided in me, in the course of a dogwatch yarn, that he was getting to the end of his tether; 'Can't do another passage with that chap', he complained, 'sits on the saloon settee and pokes around his ears with a matchstick'. I had to admit the validity of the accusation but inwardly thought the criticism might have come from a less fallible source. The fact was that, in that ship, we were frequently served with pea soup, lentil soup, bean soup and pease pudding. Which item of diet frequently caused my critical friend to break wind audibly. In a ship with an open bridge, however, this ungentlemanly lapse caused little concern to anyone. What did grate on us all just a bit was that he rarely failed to exclaim simultaneously: 'Better an empty house than a bad lodger'.

As you can well imagine, these two related incidents caused me to carry out a thoroughly searching self-analysis to identify and eradicate my own shortcomings. But as you may also well imagine I came through the analysis with flying-colours – obviously the one and only impeccably mannered gent in a ship full of behavioural visigoths!

If salvage personnel generally are peculiar, divers are undoubtedly more so. They are in a class by themselves and their yarns are horrific! They also have to be heard to be believed – or not. They will certainly tell about skeletons in long-sunk ships standing up and holding out arms in welcome; about encounters with sharks and octopi; how they were trapped upside down in a ship's hold. These stories fulfil a role parallel to that of scientists' latin phraseology; they are almost certainly intended to discourage outsiders from poking unwelcome noses into a hard-learned craft. The ruse worked well for a very long time but the offshore oil and gas activity has brought many new divers on task.

Most salvage divers are intrepid men; some might even say foolhardy. Long familiarity with the danger of the deep has made those dangers seem normal. They think little of crawling into wrecks through inadequate openings and penetrating deep down inside. They love to get somewhere or do something that their buddy has been unable to manage. They are seemingly oblivious to deteriorating weather on the surface, the tide beginning to race and the fact that they have been down an hour too long anyway. Salvage diving is one of the most demanding and frequently dangerous tasks that I have ever encountered in my quarter-century of service at sea. But in most cases – and every man-jack of them would deny this most emphatically – the divers are in love with diving.

Salvage divers need to be loyal, courageous, determined, competent and explicit. They must want to get to the task as often as possible, stay on it as long as possible, and do the maximum amount of useful work when there. They must have complete trust in their surface colleagues. I think that I have said enough to show that they mostly are, they mostly do and they mostly have.

One example of a diver's determination remains with me over the years. When I finished a two-year appointment as Fleet Boom Defence and Salvage Officer, Mediterranean, one of our finest divers surprised me when he came to say goodbye. If he had one fault it was an over-sincere devotion to the products of Scottish distilleries. I knew he had tanked up to the level of his eyebrows the night before and never expected him even to remember my departure. But at 0900 he was at my office door, stone cold sober, clean, well-dressed and immaculately groomed. The Depot Chargeman told me later that Diver 'X' had got up at 0630, run round and round the racecourse till sober, had a cold shower, black coffee and had come straight to my office. Maybe he collapsed at 1000. But – as a former soldier – he had added additional lustre to that old Army tag: 'on parade, on parade'. I'd trust him with any job I had.

It fits in with those few words about divers to say that they could achieve nothing without the people up top. Obvious, perhaps, but often overlooked when the credits are handed out. Again, the deck personnel must be knowledgeable and experienced seamen. Ships' officers must be alert to every change in conditions that might possibly lead to hazard. Despite possible disparity in rank, the men up top must be eager to meet every requirement of the men down below.

In particular again, that paragon of virtue the salvage officer must lead the way. His patience must be able to endure (and absorb without passing it on) the criticisms of the uninformed during long periods when underwater work is impracticable. And to do similarly when long periods of underwater work are themselves apparently producing no result. Any trace of induced rashness – or even irritation – will diminish the confidence of his underwater subordinates.

Harping on and on in the text about salvage officers may seem to be labouring a point that has already been well made, but they really are the lynch-pin of every operation. They are infinitely more important than the vessels or the equipment. They have been known, in the absence of any such assistance, to improvise a recovery almost with a piece of string and a bent pin. They are single-minded, dedicated enthusiasts when on the job and often when off it. It is sometimes even difficult to find another subject of easy conversation.

Some time back, I was leaning on the rail of a salvage steamer in the Mediterranean. The sun was going down in a riot of glory and colour. Sky and sea were wonder-blue. Streaks and wisps of high cirrus reflected a hundred hues of radiant beauty. Even my earthly perception realised that this was intoxicating nectar only to be drunk by the privileged few.

'Isn't it superb?' I said to my companion, a salvor of long, long experience and high repute.

'Not bad at all', he replied politely giving it a cursory glance and dismissing it as quite unworthy of further notice. 'Did I tell you of an experience I once had when working a copper cargo out of a wreck in Delagoa Bay?'.

In sum, individual members of any salvage organisation may be odd; they may be crotchety; they may temporarily fall out with one another. Being human, it would be extremely curious if they didn't. But when the strain is on, they form the most united and competent team it is possible to muster in an imperfect world. I am of Viking descent, a seafarer by choice for twenty-five years, and have shipped with many crews of many types. I'd rather be in a tight corner with a salvage crew than any others I have ever sailed with.

Chapter 3

Salvage vessels, craft and equipment

This chapter being inescapably something in the nature of a catalogue, is bound to be a tiny bit dull. But to place it later in the book or even to make it an Appendix is absolutely not on. I have too often struggled through a book with repeated consultation of other reference books only to find things made perfectly clear in the final Appendix. So I am determined to inflict no such frustrations on readers of this book. I would suggest only a brisk canter through this chapter just to know it is there and it is available for reference later. But for those who care to take their pleasures methodically and doggedly, by all means dally and hoist all the information inboard as you pass. 'Slow ahead on main engines and steady as she goes, Mister!

Salvage tugs

Any and every tug has a potential capability as a salvage tug given the right type of casualty in the right sort of environment. For instance, if the casualty is a tiny coaster aground up a narrow creek, a tiny ultra-manoeuverable water tractor may be best suited to give assistance. However, there is one point to remember. Such minor tasks will rarely qualify for a salvage award; more often than not they will be carried out by hiring appropriate assistance at a daily hire rate or – if there is little urgency involved – firms may be invited to submit competitive tenders for doing the job. Thus, special cases apart, it is generally accepted that a Salvage Tug is a large ocean-going vessel with fire-fighting equipment, a small workshop, diving facilities, pumping gear, and a certain amount of ground tackle (anchors and cables) for laying out from a grounded ship. She will, of natural right as a tug, be fitted with heavy duty winches and great towing power. In a subsequent chapter we shall have a look at the subject of towing in greater detail. Right now, we are only putting a marker down and defining the class approximately.

Salvage vessels

During World War II, and its aftermath, the proliferation of casualties needed much more assistance than plain towage. Salvage vessels evolved, therefore, with fairly minimal towing capability but which were rather more versatile than tugs. They could do impressive emergency repair work with their own stores and workshop facilities; they could lay out plenty of heavy ground tackle, refloat ships, carry out body lifts from the sea-bed and so on. But, as the wartime and post war casualties were gradually dealt with, full-time employment for them became scarcer and there are now extremely few specialist salvage vessels in service – although local wars give them additional life regularly. There is, however, one type of salvage vessel which concentrates on positioning herself over the tops of deep wrecks and recovering ships' cargo which will probably have increased phenomenally in value during its soujourn underwater.

Vessels adaptable as salvage vessels

A former Admiralty Chief Salvage Officer, Peter Flett, OBE, used somewhat impatiently to ask: 'What *is* a salvage vessel anyway?'. He would continue to state his categoric view that:

'Any vessel can be used as a salvage vessel, if it is required to do so'. He was right of course – as he invariably was in matters concerning practical salvage. Indeed, one can remember an aircraft carrier (those dear dead days when Britain possessed proper aircraft carriers!) towing a tanker to safety, extinguishing fires on the way and carrying out competent repair work on arrival at an anchorage. But one of the paramount principles of salvage – as in any other business – is to improve profitability by making the minimum outlay to produce maximum results. Therefore the use of small, handy vessels with small but adequate crews is sensible. Mooring vessels, offshore support vessels and stern-trawlers are particularly suitable for salvage work by virtue of their good working decks, heavy duty winches and good lifting capability.

Some improvisation is always necessary (it would still be so if the *Queen Elizabeth 2* with every modern convenience were to be commandeered for a salvage job). The British Navy Department cleverly did advance improvisation by designing and building MSBVs (Mooring, Salvage and Boom Vessels), to combine in one hull the main qualities of three former separate types of vessel. They can go practically anywhere and do practically anything that is required except that towing is regarded as only a bonus capability; this last function is performed from a fixed towing-hook which is 'bow-and-arrow' stuff compared to modern towage vessels towing from winches which cleverly pay out or take in wire as the towage strain increases or decreases.

Lifting craft (power lift)

This category includes floating cranes, gantry-type craft, floating sheer-legs – and vessels like MSBVs which can perform a power lift over rollers or an 'apron' in the bows. The MSBVs, as said before, can go almost anywhere, anytime. But the other types have seagoing qualities, stability and speed (either self-propelled or in tow) which make them more suitable for long-termed programmed operations than a 'rush' job. In addition, they are usually heavily engaged on tasks from which it is physically or economically impracticable to disengage quickly.

Semi-submersible deck cargo ship

A wonderful modern concept which can submerge its cargo deck in best position adjacent to a submerged casualty. At which juncture the casualty is floated, lifted, dragged or otherwise induced to sit on the cargo deck. When in suitable position, the host vessel is refloated thus bringing the casualty out of the water. When safely secured on deck, the casualty can be transported anywhere in the world for repairs. A famous recent case was when, in 1988, *Mighty Servant 1* carried the damaged destroyer *Southampton* 6000 miles from the Arabian Gulf to Portsmouth.

Lifting craft (non-power lift)

These craft usually work in pairs. In operation, the large water tanks are flooded, enormously thick steel wires are passed underneath a wreck by each lifting craft and secured on the deck of its consort. Then, by expelling water from the ballast tanks, the craft rise higher in the water and will set up the wire 'cradle' bar-tight. A repetition of the ballasting down and securing of wires will generally be required and the lifting craft will then be 'pinned down'. A subsequent de-ballasting will now lift the pair of craft with their cradled wreck so that the whole affair can be shifted into shallower water, an operation which may be repeated time and time again.

In tidal waters, the height of each lift can be greatly augmented by carefully calculated use of rising tide. Wrestling with reluctant, stiff, unwieldy wires makes this sort of operation more difficult in practice than in brief description on paper. Sixty metres of this heavy wire weighs approximately a tonne in itself and although called extra special flexible steel wire rope, the flexibility is purely relative!

Submersibles

These craft are small underwater work boats and are so versatile as to deserve their own chapter later. For the moment, it is sufficient to note that they come in three main classes.

(a) Manned: with crew in dry at atmospheric pressure; can theoretically go to any depth.
(b) Manned: with crew in wet at ambient sea pressure; restricted to depths at which unprotected crew can operate.
(c) Unmanned: directed to workplace by remote control from surface ship; can also go theoretically to any depth.

Helicopters

Although not salvage craft in the strict sense, helicopters are of enormous value in many types of salvage operation. For instance, in the initial search for a casualty, in placing a salvage officer or salvage party aboard quickly. Also for fetching and carrying fire fighting equipment, salvage pumps, stores and fuel. For passing a line from one ship to another to enable a tow-rope to be hauled across. Helicopters have also towed small craft to safety without other assistance. Although I have not seen the technique used in salvage yet, an entire ship's cargo has been discharged in helicopter trials well off the United States Eastern seaboard.

Observation chambers

Once again not truly a 'craft' but a watertight vehicle/container, occupied by an observer. A typical chamber is not too unlike a large egg-timer, so shaped for strength, and large enough for the observer to sit on a revolving bicycle type seat; it has windows so placed that the observer can get an all round view with a reasonable upward and downward angle of vision. No matter to what depth it descends, the observer remains at atmospheric pressure and no need exists for decompression drills on re-surfacing. Because the 'diver' is totally enclosed, he can do no actual work below; he can, however, visually guide a grab towards recovery of ship's cargo, aircraft wreckage, or other material on the seabed.

Similarly with submersibles, there is no theoretical limit to which a chamber can descend but there are practical difficulties. At great depths with enormous pressures being exerted, enormous structural strength (including the windows) is required; those same pressures make it very hard to ensure watertightness of electrical leads and connections to the underwater lighting which is essential as the light from the surface disappears.

Salvage pontoons

Large-sized, cylindrical pressure vessels, known by salvors as 'camels'. There are naturally differences in sizes and capabilities throughout the world. But the ones I know best – Admiralty Salvage Pontoons – will lift 80 tonnes apiece. Valves for admitting and expelling air or water are fitted externally so as to be operable by divers; when required, the compressed air is introduced by lines from a surface vessel. A wreck is normally 'cradled' between a pair, or several pairs, of pontoons which need to be most accurately and securely positioned by the divers. These particular pontoons weigh (empty of water) above 43 tonnes apiece, use large size steel wires for 'cradling' and require much exhausting and skilful underwater work by divers. Operations can therefore be extremely frustrating when there is a lop on the water, or surface swell, or in a strong tideway, or when water and air temperatures turn all fingers into thumbs. There are no merit marks for the salvage officer who inadvertently lifts the wreckage tail-heavy or bows-heavy so that it slides out of the cradle and returns post-haste to the bottom. I have personally encountered that misadventure only once but I suspect it has happened often enough around the world.

Lifting/Buoyancy Bags

Usually made of tough rubber; bags work on the same principle as pontoons. They were particularly liked by aircraft salvors in the old days when the Royal Air Force had its own salvage department and especially for giving temporary additional buoyancy for flying boats and seaplanes when the soft construction was better for the fragile hulls of aircraft than steel pontoons. They seem to have returned to favour recently, perhaps because of ease of carriage and stowage when empty. I regularly see an advertisement for stock sizes of between 0.1 and 5 tonnes with 'bags in excess of 250 tonnes to order'. The only time I had personal dealings with lifting bags was during the salvage of a cargo of mines from the minelayer *Port Napier*; it was considered safer than trying to 'sling' mines underwater to attach them individually to quite small lifting bags so that they could float to the surface and be dealt with by experts in full light of day.

Salvage equipment (general purpose)

In a strict sense, salvage equipment is hardly special equipment at all. It is a miscellaneous collection of items more usually seen in regular use for other purposes. In the seamanlike line, there are anchors and cables, slings, shackles, clamps, wire ropes (the heaviest with a breaking strain of 247 tonnes), heavy blocks, rigging screws, jacks, impact wrenches, wirecutters, grabs, cement, timber, blacksmiths' tools and shipwrights' tools. With engineering applications come ordinary cutting and welding equipment, pneumatic chippers, riveters, jack hammers, saws for ordinary and underwater use, tunnelling lances, drills, woodborers, pumps (some submersible), compressors, portable boilers, air lifts, air winches and generators.

Diving equipment

In some types of salvage case, there is a large underwater task to be done. Salvage is judged solely by the criterion of how much useful – often demanding manual – work can be carried out. Factors is selection of gear used and method preferred are: simplicity in operation, reliability, protection against cold and against denizens of the deep, ability to remain down for long periods, ability to carry out meticulously accurate work on occasions, ability to communicate freely with the surface and ability when required to 'anchor' a diver firmly in one spot by gravity.

For all these reasons, salvage divers still frequently use the heavy, old-fashioned, so-called 'standard' type of diving outfit. Men in much lighter SDDE (Surface Demand Diving Equipment) can do useful work in certain circumstances, while search and survey jobs are very suitable indeed for SCUBA (Self-Contained Underwater Breathing Apparatus). Underwater television, either lowered into position from the ship or hand-held by divers is a remarkably useful tool; it enables non-divers on top an opportunity to see at first hand what requires to be done on an underwater job or just what is going on below. Television has no magic properties and suffers the same handicaps as the ordinary human eye; it is most beneficial in clear offshore waters and least so in muddy, murky, esturial waters and rivers.

The 'ideal' salvage vessel

While lecturing to a group of newly entered members of the Royal Corps of Naval Constructors, a thunderbolt hurtled on to my head out of the blue. 'What specifications would you ask for, Commander Forsberg' asked one acute young student, 'if you could have your ideal salvage vessel built?'. *What* a question! After all the penny-pinching, making do, reducing power, cutting down of crews, compromising with lifting capability, etc., I had taken part in over the years, the offer (albeit a fancy only) to have just what I liked, was a stunner.

This ideal world-wide salvage vessel must, one feels, come in tug form nowadays. She should have tremendous towing power in order to tow a top size tanker at reasonable speed in an emergency – plus towing winches and towing wires to match. She should have really high speed to reach casualties swiftly and her own helicopter, landing platform and associated fittings; only thus can one assure speedy connections and communications in all conditions. She should have all weather diving facilities, possible down through a 'well' in the middle of the ship. She should have massive lifting capability, preferably running into thousands of tonnes.

The ideal vessel would require enormous storage tanks to carry sufficient fire-fighting foam for the worst possible maritime conflagration. She should carry her own working submersible with, again, weatherproof launch and recovery facilities. The best possible electronic navigational aids would be required to get the ship within a metre or two of a required position plus computer-operated 'thruster' units to keep the ship locked on to that required position i.e. a thoroughly reliable and meticulously accurate dynamic positioning system (DPS). Splendid workshop facilities and a commodious explosives store must be provided. Storage space for ample amounts of all that equipment previously mentioned. And, most important, we need an adequate complement of officers and men to work all the gear.

We have probably finished up with a ship of about 10,000 tonnes with a crew of 100. With all the special virtues we want built in, she will cost a mint to construct, a king's ransom to equip, and a fortune to operate – a sure-fire financial loser for her owners on even the most lucrative salvage job. Ideal operationally in fact but very non-ideal economically. So lets think again and think smaller.

Let's settle for just enough power to tow an enormous tanker in reasonable weather and merely to 'hold' her in bad weather. Let's reduce the ship in overall size, engine-room and fuel costs by accepting an economic speed of, say, 20 knots. Next cut out the helicopter, its platform, hangar and crews; communications will have to be by radio-telephone, megaphone, loud-hailer or boat. The weather is not consistently bad (we just think it is!) so we accept that diving must cease when there is wind Force five and above. Accept too that your salvage vessel will do a body-lift of a mere 200 tonnes and that only when pushed. Cut out that submersible; one will have to be flown out to you if required. Scale down the other 'ideal' requirements, especially crew numbers, and what is the answer? It is a remarkably seaworthy vessel looking uncommonly like one of the big commercial ocean-going tugs on the high seas already!

Although men in ships at sea will loathe to admit it, one thing seems to come clear from these thoughts. You know those perishing silly fools who sit on their fat backsides in head offices ashore? Looks like they do know *something* after all.

Chapter 4

Vessels aground. Refloating operations

The modern mariner has an immense cornucopia of superb navigational aids – good charts, radar, gyro-compasses, radio direction-finders, electronic depth recorders, sundry electronic, position-fixing systems (including satellites) and Ships' Inertial Navigation System (SINS). Few ships have them all but most, except the very old ones, have two or three of these aids. In addition, all ships have the traditional evergreen, navigational safeguards – sun and star sights, sounding lines, good lookouts, logs (distance recorders), charts and tide-tables. So you might well ask a very pertinent question. 'Why do ships still run aground and require refloating operations?'

There are, in fact, many reasons. Haste to arrive at a port in daylight or on a certain tide. Reluctance to lose 'face' by reducing speed when doubt is creeping in. Reluctance of a watch-officer to disturb his captain who may be snatching a tiny rest after many hours on the bridge. Steaming too close inshore to give passengers a good look at some exotic island. Mistaking one navigational mark for another. A failure in understanding between a captain and his navigational officer or pilot. Excessive economy-mindedness causing a captain to cut corners. British or American charts used by someone not thoroughly conversant with the language – or, for that matter say a Japanese chart being used by a European.

Violent weather can drive a vessel ashore even when it is 'securely' anchored. Ships can be deliberately beached to prevent them from sinking from damage incurred in an accident. A survey vessel must necessarily enter dangerous waters to work and must occasionally pay the price for so doing. Lack of training in modern navigational aids – especially in some 'flag of convenience' ships. Relying on audible fog signals, which are notoriously unsuitable for trying to estimate a range and bearing. Navigating by eye and memory in clear weather – and then unexpected fog blots everything out. Failure of ships generators, steering gear and main engines.

Let's look at just one example of how the danger of grounding may come on a ship unexpectedly then we'll leave the subject of 'going on' and change to that of 'getting off'. Due to boiler defects, the fully loaded cargo ship *Steel Vendor* (7,752 tons) lost all propulsive power on 5 October 1971 in the South China Sea. Because of bad overcast weather she had been unable to fix her geographical position for days. She wallowed helpless in heavy seas and hurricane-force winds. At last, a hazy sun appeared on 7 October and, while officers were quickly snatching sun sights, white water was seen breaking on the port side horizon. It was a reef and the vessel proved to be 90 miles south of her previously estimated position.

The Master calculated his drift and realised that he must inevitably hit the reef quite soon if the main engine could not be activated; he informed the chief engineer of the need for extreme haste. Three hours of mental agony followed, an anchor held the ship for ten minutes before its grip was overwhelmed by wind and swell. The Chief Engineer, was at long last, able to tell the Master that limited engine power was available but the ship struck fatally almost at that same moment. So far as ship's personnel were concerned that incident had a happy ending; helicopters from nearby HMS *Eagle* lifted them all to safety. But it is just one random example of the unexpected grounding and loss of a good ship.

There is a useful lesson to be learned from that case; a lesson that new generations of men have to learn painfully time and time again. It is that weather can so often be the supreme ruler of ships' fates and mens' fates. Landlubbers will ask why the ship's engineers could not hurry up their relatively simple repairs or at least improvise something. As a matter of fact, so might some marine engineers who have come ashore as advisers and have been sitting in steady armchairs on a firm office floor too long. Will all those questioners please note this extract from the US Coast Guard Marine Board of Investigation Report.

> The situation aboard the *Steel Vendor* was growing grim. She was dead in the water, rolling 35 to 40 degrees in heavy seas and Force nine winds. Bilge water was about 14 inches deep and was sloshing over the deck plates in the engine room. Only limited power was available from the emergency generator for lighting and ventilation. Hand tools were sliding across the deck plates and disappearing in the bilges. Sea water was coming down. the stack [funnel] from the wind-driven seas and was dripping on the men and equipment.

All that ghastliness plus the knowledge that the ship was inexorably drifting to destruction and to probable loss of life. No wonder only real men can stay at sea for a lifetime. Coal miners ashore may get danger money but there is as much danger afloat and the extra payment is zero.

We have just observed a stranding on rock – albeit a rather special case. But, equally well, strandings can occur on sand, shingle or mud – or even on occasions, high in a paddy field. In that last category, I somehow always bring to mind the *Clan Alpine* which was innocently at anchor at the mouth of the Karnaphuli River awaiting allocation of a harbour berth. On that night, of 31 October 1960, Chittagong and vicinity was devastated by a cyclone with 120-knot winds. The *Clan*

Alpine was driven eight miles by the wind before finishing in her paddy-field. Next daylight coincided with time of low water and the vessel was found to be a half-mile from the nearest sea. A road was eventually built to the ship, all cargo discharged into road transport by the ship's own derricks and the ship sold locally 'as lying'. As a memento of that meteorological upheaval, a new watermark was said to be clearly defined on a local lighthouse – 11 metres above normal highwater mark.

As already said, every single salvage case is a law unto itself. But there is one golden rule in respect to strandings. If you have just gone aground then, with luck, there is an outside chance that you might equally just come off on the same tide or the next. Weather permitting, it is often well worth a quick try within twelve hours. So radio for tugs if they can get there in time. Lay out to seaward the heaviest anchor that can be man-handled in the time available and bring the anchor wire to a winch. Lighten the ship by getting rid of some water ballast. If only one end is aground, maybe a transfer of water ballast from one end to the other would trim (tilt lengthwise) the ship into a more floatable attitude. That's the scene setting; now for the play itself.

About two hours before high water, get the tugs to take the strain on all towing wires. Take a gentle strain on the anchor wire; too much strain may merely 'bring the anchor home' to the ship instead of vice versa. Put the engine at slow-astern; putting it full astern for a sustained period may overheat the bearings and may also aerate water around the propeller and cause it to lose grip. With all these forces applied, the last hour of rising tide may well see the vessel commence a slow movement to seaward. But shortly before the top of calculated high water everything must be operated at a maximum practicable power for the situation is now critical. (In my training ship, we used to practise 'roll-ship' by having the whole ship's company charge *en masse* from one side to the other in order to break the bottom suction. But ships since got larger and crews smaller!) If she comes off, that's fine. If not, then the 'quick pluck' has failed and a more prolonged, carefully planned, refloating operation is going to be necessary.

Just sometimes, one is even luckier than having to await the next tide. I remember coming down the River Plate, accurately in the correct navigation channel when we took the ground. The mud was so soft, and our speed so moderate in restricted waters, that no one noticed the impact. It was some moments before we realised that we were no longer passing the nearby navigational marks! However, on a rising tide, we managed to back off fairly quickly, select a slightly different part of the channel and bust over the mud bar into deeper water. There are hundreds and thousands of cases like that and notation of the loss of professional dignity is marked only by internal disciplinary action. They are the lucky ones who escape the headlines.

Very occasionally, and entirely reprehensibly, such temporary groundings are not reported on the principle of 'least said soonest mended'. I have vivid memories of seeing in drydock a destroyer whose captain maintained that she had never even 'smelt' the ground. Very curious how there were great dents in the bottom, extensive longitudinal grooving and large areas of paint missing! Could have hit a gigantic whale of course – except that it was a Mediterranean destroyer where gigantic whales are in short supply. Earlier than that I had been shipmates with a Second Officer who confessed to me that he had lightly bumped the ship on an extremity of the Goodwins one night; he had also gone on his way very silently rejoicing at his undeserved good fortune.

As we have just dealt with immediate refloating from a soft bottom, let's go on to a sustained and planned operation from the same situation, that is, of stranding on mud, sand or shingle. Although planned, that does not mean that the operation is carried out in slow time. All salvage operations proceed as fast as possible consistent with the acquisition of the necessary resources. Sitting on the bottom – even the softest of bottoms – subject to tide and wave action, never does any good to the structure of any ship. So the less time in hazard the better.

I can confirm that last point from bitter personal experience. I was in command of a 1,000-tonne steamer refitting in a small yard with only a mud berth available. For two months the bows were firmly aground on mud while the stern rose and fell four metres on each tide. At the end of the two months, she was refitted beautifully, came off the mud sweetly and looked like a yacht. But either the hull had become distorted with the twice-daily bending stress or the shafting, between engine and propeller, had suffered likewise. Despite repeated visits to repair yards, and repeated assurance that all was well, the ship was never fully operational again. Time and time again the Chief Engineer came to the bridge, almost in tears, to report the bearings running hot and to request a reduction in speed. That unfortunate vessel finished her short life at between half and three-quarters speed – and was sent prematurely to her ultimate end at the breakers yard.

So you're stuck hard on the putty (our omnibus term for any non-rocky bottom) and want to come off? As the soldiers say:'Time spent in reconnaisance is never wasted'. So appoint one good officer in a boat to take soundings and then to adjust those depths, necessarily taken at different times, to one common level; salvage officers differ in opinion but I prefer Mean High Water Ordinary Spring Tides (MHWS). From that universally understood tide level, any required tidal calculations can easily be made.

In the Ministry of Defence we sometimes cheated by sending 'raw' figures to the Hydrographer of the Navy; reduction of soundings to a common datum is the strong ale and red meat on which the Hydrographer's staff thrives. Be that as it may, a quick and careful survey will reveal the best 'escape route' for the grounded casualty. Simultaneously with the survey, another officer (generally the Chief Officer) will have worked out exactly what depth of water is required for the ship to float in her present state of loading and ballasting. By subtraction of the calculated highwater level from the minimum floatable depth, you now know how much the ship is 'short of her water'.

From now on, salvage proceeds by pulley-hauley, strong-arm methods, application of a few physical principles and reductions in the ship's weight. Nothing in these techniques is black magic and all is plain commonsense. Some critics might call it 'brute strength and ignorance'. Just ignore them!

Next step is to lay out ground tackle in the best practical direction. This direction is not always the same as the direction of maximum efficiency. For one instance, the salvage officer gets zero marks for popularity if he lays his tackle across an underwater power cable, or telephone cable. (Remind me to tell you later – very much later indeed – how I once cut all telephonic communication between an offshore Army fort and its shore headquarters). And, again, unless the salvage officer is brimming with confidence, the tackle is not usually laid directly towards the best escape route; it is customarily angled off to one side or both sides. This offset procedure enables one to use the ship herself as a lever and the position she is aground as a fulcrum. Remember Archimedes – 'Give me a fulcrum and I'll move the world' Well! It's exactly like that in salvage of stranded ships.

If there is a vessel of 200 metres in length, hard aground 50 metres from one end, the vessel constitutes a very efficient lever in herself. Naturally one brings the wires leading to the ground tackle to the strongest available winches on board. Heave the wires taut and leave them for the time being. The catenary of the heavy wire and chain will exert a constant pull on the ship. And, wonder of wonders, the ship sometimes even creeps towards the anchor and comes free.

One of my great friends, and early salvage mentors, Victor Campbell, had a remarkable success on those line. With a large British merchant ship aground on the North African coast near Cape Bon, he had no immediate means of lightening her sufficiently. He therefore set up his ground tackle bartaut on the ship's powerful windlass and left steam on the windlass with the throttle open. He 'smelt' the approach of a strong offshore wind and, although that might well be desperately bad news in the end, it could just give him a favourable hour or so before its arrival. Which is exactly the way it happened. In advance of the wind, more water was blown on to the shore and a rising swell started to run in. Vic said the atmosphere was electric on board as the windlass of its own accord, started to creak, groan and heave in an inch at a time. The watching officers thought it might possibly be the anchors 'coming home' to the ship. They became jubilant when a slight tremor of liveliness showed that it must be the ship, moving almost imperceptibly to seaward. The windlass was then tended with loving care and the casualty came afloat in time to make her ready to withstand the effects of the onshore gale.

Incidentally, on mud and soft sand, salvors have another short cut which could be well worth trying. One or more tugs thoughtfully positioned, accurately trimmed and secured alongside can work propellers and scour away tremendous quantities of bottom material from around the casualty. Captain Pat Polland, one time Deputy Director of the Admiralty Salvage Department and another of my early tutors, told me he once got a depth increase of 2·5 metres in twenty-four hours by this method. Harbour and estuarial authorities are well aware that such redistribution may adversely affect their dredged channels. But they can always dredge again and, on the whole, that is a lesser evil than having a few thousand tonnes of undredgeable ironmongery lodged on their front door step.

After the short cuts have failed, the time has come to throw every artifice into the battle to get the ship nearer her water. Lighters and barges and coasters are needed to take off cargo; in the case of tankers another tanker may be required alongside to take off a few thousand tonnes. Fuel and stores may have to be removed. Sometimes masts, funnels and superstructures may have to be removed. Every tonne counts. It is heartrending to attack a fine ship like that. But desperate situations demand desperate remedies and we can't afford to be squeamish. Every vessel carries a table in the ship's documents of 'tons per inch immersion' or 'tonnes per centimetre immersion'. They tell relentlessly the number of tons/tonnes to be discarded to get the ship to float in one inch/centimetre less depth of water.

When, by every necessary measure, it finally looks as if the ship will get her water, the *moment critique* is imminent. If tugs are available, their tow wires must be well secured. Await the approach of high water. Tend the winches and take maximum strain on the ground tackle. Work the ship's own engine. If calculations are correct, she'll come sweetly off. If calculations are wrong, lighten her a bit more, wait for a higher tide and try, try again. Waggle one way, waggle the other, trim the ballast, wrench and pull. *Moments critiques* can come several times in our business!

The analogy to the refloating technique is that of a spike driven into the ground until it seems immovable. But hit it one way and then the other. Use the spike as a lever and then apply the force where it has most effect. Very gradually the hole will be enlarged, friction reduced and ultimately a direct jerk will bring it free.

In all I have written so far, it has been assumed that the vessel is surrounded by some water at all states of the tide. That is of course, not always the case as we have seen from the facts of the *Clan Alpine*: the casualty may be high and dauntingly dry. In these cases a channel to the water can be dug out by bulldozers or sluiced out by high pressure hoses. Rollers and greased boards may be fashioned into a sort of slipway. Sand, mud and earth can be jetted away to allow insertion of boards under the ship. Every device to break the suction and lessen the friction is valuable. Offshore ground tackle may be connected to the ship through enormous eight-fold purchases to give hundreds of tonnes of pull. There is no doubt that practically every stranded vessel could be refloated but money must be the supreme arbiter. If it costs less to write off a casualty than to recover her, so be it. Rest in peace. No use chucking good money after bad. The underwriters pay out on a 'constructive total loss', assume ownership and the ship is frequently sold for breaking up on the spot.

So much for soft-bottom strandings. Now for rock. Many of the same techniques apply. Surveys, tidal calculations, weight considerations, ground tackle, tugs, wrench-and-waggle. However, one point requires great attention. If the first quick attempt at refloating fails, it is generally the practice to keep the ship as motionless as possible until conditions are favourable for the next attempt. This is because a 'lively' ship can wound herself mortally by pounding and thumping on a rock bottom. Keeping the vessel still is usually achieved by tugs or offshore anchors connected to the seaward end of the stranded vessel, by pumping in water ballast and by transfer of cargo or fuel to keep the vessel firmly down on the contact area. Of course, weighing her down may pierce the hull again but such damage is generally preferable to the rending and tearing of a ship, surging in an incoming swell. Just one more expensive decision for the salvage officer to make!

Instead of going through hypothetical cases, I think refloating from rock will best be illustrated by two actual cases some fifty years apart in time. Both are interesting cases in themselves. But more than that, they demonstrate very well that although tools change with the passage of time, basic techniques do not.

The first case is the grounding of Alfred Holt's *Cyclops* on a westerly projection of Shi'b al Kabir off Jeddah in the Red Sea. It occurred at 0458 on 7 February 1910. My father-in-law, Dr. F. W. Hogarth, was the surgeon and I quote extracts from his notes:

> The crash was terrific as several thousand tons of ship and cargo reduced speed from 13 knots to zero in a couple of seconds. The wash which had been following the ship now overtook her and broke heavily on the reef: the consequent noise prevented speech for some little time.
>
> When daylight broke, there was 10 feet depth at the bows and 96 feet at the stern. The bows had almost climbed out of the water with the forepeak and No 1 hold torn open to the sea. No. 2 hold seemed tight which was a reassuring thought against the possibility of the ship slipping off. All hands worked (including the surgeon) to shift the cargo of whisky, brandy and soap out of No. 1 'tween deck to enable our carpenter to get to work on shoring up and tomming down. No. 2 hold then turned out

to be leaking: the double bottom tanks beneath it were full. The whole situation was extremely serious.

9 February. The carpenter had made his best efforts in shoring up No. 1 hatchway etc., and a lot of cargo had been shifted aft. It was assumed that 800 tons of pig-iron from the forepeak would drop out on to the reef. So Captain Hazeland put the engine full astern. To every one's chagrin nothing else happened. Then a kedge anchor was laid out astern and a winch laid on, but again she never moved an inch.

10 February. We had a nightmare of cargo shifting.

17 February. For the past week we have been getting cargo out of No. 1 hold, throwing lots of whisky overboard. The small salvage steamer *Mayun* had come to us from Perim and the *Hermes* (Swedish) from Syracuse, the latter much better equipped than the former. Then HMAS *Edgar* who was passing on passage to Aden, arrived. She got exactly astern; her thickest wire was gradually and tightly stretched, when Alfred Holt's bitts gave up the struggle and flew out to sea. But *Cyclops* never moved. HMS *Edgar* recovered her wire, gave us a toot and stood away down south. Ships of our company, *Laertes* and *Dardanus*, also called in passing. More exciting exhibitions of snapping hawsers, wires and bitts but *Cyclops* won't move.

22 February. For several days now, salvage divers have been planting shots of dynamite to shift the coral which had come through into the ship. Each day the Captain hove on two hauling-off wires and had a try with engine full astern. Today the ship floated quietly off.

23 February at sea. Ship steamed to Jeddah with pumps going. The salvors make a watertight patch, as large as the floor of an ordinary room, made of floorboards and covered with many layers of canvas, the outer edges of which are rolled into a cushion right round the patch. It is placed in position, improved by divers, and guided onto bolts brought from inside the hull and made completely watertight. Finally 30 bolts and three wires hold the patch firm. The tanks below the damaged hold are filled with concrete – 50 casks of cement and 75 tons of sand.

Work done had included throwing overboard – 800 tons of pig iron, 125 tons corrugated iron, 50 tons tin plates, 35 tons of washing soda, 2 tons of buckets, 10 tons of Gossage's soap, 1 ton of Muntz metal, 30 tons of iron bars, rivets, etc., and 15 tons of machinery. About 400 cases of whisky were smashed, as well as cases of brandy, equal to 20 tons, hence the occasional drunks. However a total of 270 tons was carried by the crew into the afterholds. A further 320 tons of general cargo was shifted into the after-holds as well.

20 March. Pronounced seaworthy. Sailed for Penang at 10 knots, the wires over the patch vibrating with the rush of sea through them: it sounds like a cotton mill.

So there we are – a 1910 salvage case lasting six weeks. Techniques exactly the same as today. We might use a helicopter to put the salvage officer aboard and to transport personnel and cargo ashore. We would have used plastic explosive instead of dynamite. We might by these means have got the ship off a few days earlier. But that's all. Nothing else has changed.

For the next case, I am indebted to Commander Thomas N. Blockwick, USN, formerly Ship Salvage Officer, US Seventh Fleet. He described the case excellently in the *United States Naval Proceedings* and the following is a precis. I have selected it because it deals with a very out-of-the-ordinary rescue of a ship from destruction.

On 28 April 1962, the Panamanian SS *Dona Ourania* grounded on Pocklington Reef about 200 miles south of Guadalcanal. The reef is submerged except for a small spit of sand about the size of a football pitch at the northern end. The reef is about 25 miles long and some miles wide. It is about 200 miles from land and is not readily visible in even the best weather. The *Dona Ourania* was steaming at 14 knots when she struck and (unlike the *Cyclops*) there was a barely discernable deceleration and some were not aware that the ship was aground. After initial unsuccessful attempts to back off, all personnel left in a coastal ship.

The USN salvage ship *Bolster* called at Guam to embark the Salvage Officer and arrived at the casualty on 15 May. The ship was a 12,000 ton grain carrier, 483 feet long with a beam of 62 feet. *Bolster*'s inspection showed that she was 2000 tons aground over the forward 180 feet. After that the depth of water dropped precipitously away to 500 feet. The ship was empty of cargo and had insufficient water ballast for lightening; the water depth aft was too deep for ground tackle. The most practicable procedure was to use explosives on the coral and to use the *Bolster* as a tug. The tidal rise and fall was only one foot; and that once on a twenty-four-hour instead of the more normal twelve-hour cycle.

Trials with explosives were successfully carried out and on 16 and 17 May the first charges were fired alongside and under the ship. They too were successful and the ground reaction point shifted 20 feet forward. Charges were placed in coral caves by SCUBA divers. In view of the now optimistic outlook a nucleus of the ship's crew were asked to return and activate the main and auxiliary machinery. This was achieved by 19 May.

On 23 and 24 May, more shots were fired and the point of reaction shifted another 30 feet forward and the ship was now estimated to be only 1000 tons aground. *Bolster* attached a tow wire to the stem and wrenched the *Dona Ourania* through 70 degrees. Eight more underwater shots were fired. On 26 May, *Bolster* hooked up to the stern and wrenched it 11 times through an angle of about 60 degrees with the *Dona Ourania* backing full. On 27 May, wrenching commenced again – 60 degrees to 90 degrees every time. Each wrench took about 15 minutes and moved the ship 2 feet seaward.

Finally the ship came afloat, steamed to Brisbane under her own power and then to Japan for permanent repairs. She subsequently went on for a long and successful career.

That case was specially mentioned for two reasons. First, because it was a seemingly impossible casualty in a remote oceanic situation with the minimum external assistance. It was carefully thought out and executed and must be regarded as a model operation. Second, it was probably the first time that blasting was used as a primary and not a secondary means of refloating.

Now to one more permutation of the refloating business. It is occasionally more economical to salvage part of a ship rather than to waste too much time on recovering the whole. There are several good examples. But the only one I was personally connected with (and that only slightly in the final stages) was the Swiss motorship *Nyon* hard aground on St. Abbs Head, Scotland, in 1959.

The usual methods having been unsuccessful, the salvors decided to abandon the bows portion and to go for the valuable stern portion. The first requirement was for stiffening up and making watertight the section to be recovered so that it would, in its own right, be reasonably buoyant for a short sea passage. Then cutting operations began and progressed well in both in air and water. The final division was expected to be made by Mother Nature by way of fracture of the keel plate after a great number of tidal ups and downs. But there's no one so cussed as Ma N. when *invited* to assist! The final severance had to be done by a naval party with explosives. The stern section was then towed to the Tyne for inspection and on to Rotterdam for a new bows section to be built on.

I have just written of a very successful salvage operation by the US Navy. It would not therefore be unkind to look back well over half a century to one of the less triumphant episodes. It resulted from the grounding of the submarine H-3 on an exposed sand spit near Humboldt Bay, North California in December 1916. I am not poking any Charlie at the USN; no salvor would ever do that to another. In our business there is always much sympathy and a feeling that; 'There, but for the grace of God, go I'.

When H-3 grounded, the first refloating attempt by a Navy tug failed and she returned to base. Then commercial firms were invited to tender and two bids were received. A big and reputable firm asked for $150,000. The other bidder, a small lumber company asked for $18,000 and proposed to transport the submarine across the half mile wide sand spit and to re-launch it in calm water on the non-exposed side. One bid was rejected as too high; the other as too unrealistic. So the Navy was ordered back to the job.

Almost unbelievably the *Milwaukee*, a first class cruiser, 10,000 tons and eleven-years old took over the salvage task-commanded as a temporary appointment by a lieutenant with no more than ten years service since leaving the Naval Academy. One man was lost and a whole boat's crew were hospitalised while getting a line ashore. Finally, *Milwaukee* got a 3,000-feet tow line connected and started to pull. The longshore current caught her broadside on and before the towing wire could be slipped, she was in the surf herself and bumping to her last resting place. She was eventually sold and broken up on the spot.

Before leaving that treacherous beach, however, let us note what happened to the original casualty, the H-3. That small lumber company did lay a track across the width of the sand pit; they did trundle the submarine across on rollers; they did launch her again in the sheltered waters of Humboldt Bay! The submarine was then refitted and served her part in World War I. More thought and less impetuosity would have saved the loss of a first class cruiser. But that is – as they say – how the cookie crumbles.

Because refloatings after stranding make up a great proportion of the salvor's business, this chapter has been longer than average. To be sure of overall coverage, hundreds of different case-histories ought to be dissected and examined in detail. But – to coin an especially unique phrase – time marches on! Let's have a look at one seamanlike piece of good fortune (there's not enough about to ignore when it happens) and then to move to another subject.

The *Seiko Maru No. 8*, a Japanese tuna long-liner ran aground on Frederick Reef, 290 miles off Gladstone, Queensland, on 14 July 1968. The crew abandoned her as a hopeless case and were taken ashore by another tuna boat. After nearly a year of the vessel taking all that the weather could hand out, Captain Henry Renton, in the tug *Yarwum*, sailed from Gladstone and returned one short week later with the *Seiko Maru No. 8* in tow! The customary bit of pulling and slewing had done the trick again. From the gleam-in-the-eye stage to completion, it had taken only four weeks; the operation cost only $A12,000 which is no more than pin money in the salvage world.

One big drawback was that the fish catch had, not unnaturally, gone rotten and one man was even rendered unconscious by fumes. Despite being unofficially named *Stinky Maru* by the salvage men, the sweet smell of success must have been doubly enhanced as they steamed proudly into harbour dragging an 'impossible' salvage case behind them.

That, also, is how the cookie sometimes crumbles! But not very often.

Chapter 5

Sunken vessels. Raising. Refloating. Removing

When a model boat sinks in a boating pool, the owner goes to its assistance, lifts it, empties it out, examines it for damage, carries out any necessary repairs and refloats the boat. That, in essence, is what any salvage officer has to do in any ship-refloating operation. The main difference is that, because ships weigh thousands of tonnes, he has to think up some fairly ingenious methods of lifting and emptying. Basically, the methods are few but the variants are many.

First, if the casualty is hard on the bottom in fairly shallow water, the decks and deck openings may still be above the water. In which case, it is simple to shut any underwater openings and patch any underwater damage by which water initially entered; it is then possible to regain buoyancy by pumping the water out. When I say 'simple', everything is of course relative! If water is removed and no more (or little more) water can enter, the ship will ultimately come afloat in a manner and attitude determined by which compartments the salvage officer chooses to pump out first.

A variation of this method is when the deck openings are covered at high water but are workable at low water. This situation requires accurate replanning, assembly of a great mass of pumping equipment and an all-out onslaught to regain buoyancy between tides. Emptying speed may be achieved by cutting a large hole in the ship's side to let out water on the falling tide and patching that hole over again before the rising tide reaches it.

One further variant in this method of refloating is when the deck of the casualty is frustratingly close to the surface but never comes sufficiently clear of the water to work on it. The procedure now is for divers to seal up every opening in the hull except selected ones, usually hatchways. Those hatchways are then built upwards with steel plating or timber in a sort of tower effect until the open top of the structure comes well clear of the water. Pumping can now be carried out as before and the ship will come afloat providing the underwater sealing of holes and openings has been effective.

There are, however, two vital aspects. The towers (or cofferdams as they are properly called) have to be built pretty solidly and to be well supported by strong stays to withstand ordinary wave surge and tidal pressures. Further, the operation must be pushed forward relentlessly, without pause, because no temporary structure can hope to withstand really heavy battering from the sea in gale conditions. Many salvors' hearts have been broken as a result of the collapse of a cofferdam after months of hard labour.

A second distinct type of operation arises when the ship is sunk too deep for cofferdamming. Such a casualty can be lifted physically by tidal lifting craft, by floating-crane or by powered gantry-type lifting craft. This is a splendid and often self-sufficient method for small ships but it is impossible to apply enough lifting power to deal with large sunken vessels. With the latter, sufficient buoyancy must be recovered, by use of compressed air, to make the weight of the ship liftable.

Good stuff, compressed air. But it needs watching. At one moment it may be applying all the right forces in all the right places. it may require, however, the attitude of the casualty to change only slightly and all hell is let loose as the air applies all the wrong forces in all the wrong places. As just one example of this, Captain W.R. Fell told me the awful now you see-me-now-you-don't-story of the Malta floating dock. I can't pass it on better than he later wrote it in *The Sea Surrenders*.

> Things had not gone to plan, and the first thing we saw on entering harbour was this 12,000 tons section [of floating dock] poised like a ballet-dancer on one toe. I arrived off the dock just in time to see it give a vast drunken lurch and, midst a seething boiling mass of air bubbles, try to roll over, hit the bottom before it could do so and settle back on the seabed ... . The section had buoyancy but no stability – the moment it left the bottom it tried to capsize.

That short paragraph graphically describes the instant of failure. But it does not tell the months of planning, surveying, diving, patching, welding, testing and finally blowing. In a relentlessly cruel maritime game of snakes and ladders, the salvors had slipped back from very near the end of the game right back to the very beginning. Disappointments like that are part of the salvage business. There's nowt to do except to heed King Robert the Bruce, pick oneself up off the deck and try again.

Instead of using compressed air, some modern salvors have favoured injection of polystyrene. This is a chemical looking like granulated sugar; when steam-heated, the granules expand up to fifty times their original size. Self-evidently, as the substance is injected, it pushes water out of the vessel and, additionally, has a certain amount of buoyancy itself. The operators estimate that, give or take a bit, successful injection can be made down to depths of 75 metres before mechanical and physical difficulties make it over-difficult.

The main advantage of polystyrene is that it is much less likely than compressed air to escape from a right compartment to a wrong compartment. Secondary advantage is that it pushes only upwards towards the sea surface instead of, as compressed air does, in all directions. And finally, it remains very much at a constant volume despite changes in ambient water pressure. On the other hand the primary advantage of air is that there is plenty to be had for free in any part of the world. And, having done its job, it just fades away with no disposal problems!

One word of warning is necessary about both agents. Ships are built to stand great pressures from outside and are naturally never designed to stand anything like the same pressure from inside. So, before applying internal pressure, it will probably be necessary to strengthen and stiffen the outside by strategically placed beams and girders. And in the case of compressed air, never, never forget that clever Mr Boyle and his Law, that is, if temperature is constant, 'the volume of a given gas is inversely proportional to the pressure'.

In practical terms, say you have a compartment full of air at 10 metres depth which gives a pressure of 2 atmospheres. Then bring that compartment to the surface, where there is a pressure of 1 atmosphere, and that air will want twice the space to contain it. Try to keep the doubled volume of air in the same compartment, and there will be trouble. Very interesting trouble, but not a bit of conducive to timely and successful refloating. It goes almost without saying that expansion of air is even more troublesome when raising a ship from a depth equivalent to 3 or 4 or more atmospheres. The remedy for this expansion of air is to ensure a slow rate of ascent for the ship and for an efficient venting system for excess pressure.

In the early post-war days, when wrecks were plentiful, and scrap metal was in great demand, a good many 'instant' salvage firms appeared on the scene. In a bid for maximum turnover and profitability, they relied heavy on compressed air and on hoping for the best. Naturally, they had some spectacular, albeit nail-biting successes – and often they had even more spectacular failures. A sunken ship would leap to the surface in a miniature tidal wave, have one final moment of proud glory on the surface before the compressed air burst out and the ship then disappear ignominiously in a flurry of foam. Sometimes for ever. In those times there were enough wrecks to abandon the difficult ones and move on to the easy ones. Happy days!

Before leaving the theory of the subject, it might be as well to say that it is usually (but not always) most convenient to work with the casualty in an upright position so that something like normal conditions prevail. It is therefore pretty usual, if a vessel is sunk on her side to roll (the correct term is 'parbuckle') her upright.

The ship's centre of gravity (CG) can always be calculated fairly accurately. So some wires are attached to pull one way below the CG; others are attached to pull the other way above the CG. Pylon-type structures may be welded or bolted to the ship's side for the high-side wires to run over and thereby achieve a lever effect. Weights may be hung on the high side and external flotation devices attached to the lower. Some compressed air may be injected to lessen the dead weight of tonnage to be rolled over. It all sounds easy but can be extremely difficult when actually on a job.

Now then, let's go from that easy theory to difficult practice by looking at one or two actual cases. Lesson number one is perhaps that small ships do not always make small salvage jobs. For instance, the small Dutch coaster *Sertan* (850 tons deadweight) sank in the Manchester Ship Canal on 14 November 1960. As the sinking position was fortunately clear of the main traffic channel, the Manchester Ship Canal Company initially permitted the Dutch owners to engage their own salvors from Holland but they did not begin work until mid-January 1961. A small ship in sheltered water, close to every kind of logistic support – couldn't be easier one might think! But it was not until mid-March 1963, after fourteen months of work, that she was finally cleared.

The situation facing the salvors was that *Sertan* had been secured alongside a wooden dolphin while undergoing a search for underwater damage sustained during her approach to the canal entrance. It appeared that, after holing herself on a stone slab, she had unwittingly been kept afloat by a large amount of air trapped under the steel hatch covers. When these covers finally succumbed to the pressure (in exactly the opposite direction for which they were designed) the trapped air bubble escaped and the vessel rapidly listed towards the dolphin and sank.

The incoming Dutch salvors planned to remove cargo, regain as much buoyancy as possible, bring her upright by a combination of heaving her up the slope of the canal bed and parbuckling. Then to review the situation in order to proceed as the new situation demanded. Nothing much wrong with that plan – couldn't have done better myself! But with the unutterable cussedness of some salvage cases, the plan didn't work out.

Mishaps began early. First an attendant vessel broke a mast lifting too heavy a weight. Then it proved more difficult than anticipated to find 'secure' anchorage points ashore for the parbuckling wires. But still reasonable progress was made. More than half the cargo was removed, most of the double bottom tanks were emptied of water and the top weight of both masts had been cut off. Thence to the first righting attempt; it failed. At that point, lifting pontoons were introduced to increase the turning moment at the same time as lifting bodily; the second attempt failed. In September 1961, came partial success and the vessel settled at a small angle of heel of 10 degrees.

Now, with a certain amount of buoyancy achieved, the salvors decided to heave the ship astern in the belief that she would slide and ride up the bottom silt to come completely upright. Alas! She dug in instead of sliding up. So they attacked the engineroom (which was reasonably watertight) with pumps in order to lighten the ship further. On 2 October, the final Dutch uprighting attempts was made; the ship became temporarily buoyant, then rolled over and sank in almost precisely the same attitude as she had been at the beginning of the ten-months work. It was a desperate situation.

Although it simply must have happened a good few times over the years, I have never personally known a Dutch salvor take his teeth out of an unfinished salvage job before. But this one did exactly that. Very wisely, he decided to cut his financial losses; he abandoned the operation and returned to Holland.

The Manchester Ship Canal Company, now left with full

responsibility for an awkward and unwanted wreck, put the case in the hands of the Liverpool and Glasgow Salvage Association – a concern with a long and proud record of active salvage since before World War I. The salvage officer appointed in charge was Captain A. Lyness, already professionally well-known and later to be acknowledged as one of the best in the country. Things began to move again.

Silt was removed by airlifts and ejectors. More cargo was taken out. The steel hatch covers were bolted into place by divers. After a nightmare search in muddy water, entirely by feel, the original damaged hole was found and patched. Two large Admiralty salvage pontoons were borrowed and connected under the hull to the high side; they provide a substantial 160-ton lift as well as a turning moment. Silt was scooped away so that the pontoons could initially sink as far as possible into a trough, in order to have a longer pull before getting to the surface and becoming ineffective.

With the hold now pumped dry, and with the lifting of the pontoons, the ship again came alive and took off 19 degrees of list. Now, with the use of some of the former Dutch salvor's parbuckling concepts, *Sertan* came to within 18 degrees of upright and finally almost on to an even keel. The ship was not only filthy with silt but found to be heavily weighed down with it. Even after concentrated hosing of tons of silt material from accessible exterior surfaces, there was an estimated residual 150 tons in various compartments. It was this astonishing accumulation which had so greatly confounded lifting and turning calculations.

In addition to points of technical detail, there is an important salvage principle on show here. For maximum profitability and maximum professional satisfaction, the salvage officer should deploy the absolute minimum of resources consistent with prompt success. But such an ideal situation, because of the many imponderables, can only rarely be achieved. It is therefore always better grossly to overestimate the forces required than even fractionally to underestimate them. 'A costly success is always preferable to a costly failure'. That is Forsberg's Law – conceived and enunciated at this very moment!

The next case is that of a bigger ship in the River Thames. I must state categorically at the outset that I was in no way concerned with this operation. But every time a salvage casualty occurs in the near vicinity, one makes a provisional assessment and thinks through a provisional plan 'just in case'! That's human nature. I was Assistant Director (Salvage) at the then Admiralty when the *Magdeburg* sank, I knew the river-staff and it was easy to see the vessel from time to time and observe progress. Industrial espionage? Not at all, for there are no secrets in salvage principles the world over; it is however vastly educative to see competent practitioners from other countries doing their own thing in their own way.

The *Magdeburg* had been in collision with the *Yamashiro Maru* off Broadness Point on 27 October, 1964. In a very short time *Magdeburg* sank on her damaged starboard side. Her cargo was 1,900 tons of general cargo and forty-two omnibuses for Cuba (which consignment gave rise to years of rumour that the collision was engineered by enemies of Cuba). Fortunately the vessel had managed to get clear of the main fairway before sinking. Nevertheless the Port of London Authority was swift to try and secure her there in addition to laying out wreckmarking buoys.

The *Magdeburg* was a vessel of 10,700 tons deadweight. As she was of East German ownership, East German salvors were permitted to work in the Thames because the Port of London Authority's Salvage Section was fully tasked elsewhere. The parbuckling upright was done somewhat uniquely by getting two enormous floating sheer-legs to pull vertically upwards instead of the more normal horizontal pulls. The reason for this procedure was in part due to the type of resources available to the salvors; in part it was to avoid the necessity to lay horizontal parbuckling wires across the main shipping channel. In due course, on 7/8 July 1965, the vessel was successfully parbuckled to 20 degrees off the vertical; the slope of the river bed precluded anything better than this. Then followed the enormous job of patching, repairing and making the vessel ready for sea.

As ever, an abbreviated account cannot possibly tell of all the problems of weather and tides and the sheer damned reluctance of very heavy material to be coaxed into the right places. Or the frustrations of the first abortive righting attempt, of delays, of legal negotiations, etc. Suffice it here to report the most awful frustration of the whole boiling issue. Four days after triumphantly leaving the Thames in tow, the *Magdeburg* sank in the Bay of Biscay – this time for ever. Nine months blood, sweat, toil and tears had all been expended for nothing. The main bonus was therefore not to the insurers but to the Port of London Authority. At least the *Magdeburg* was now sunk in someone else's bailiwick.

One splendid example of well-planned refloating was that of the Canadian Pacific liner *Empress of Canada*. This vessel was given an overdose of water during firefighting operations (on 25 January, 1953) rolled over, admitted more water through various openings and sank on her side in Gladstone Dock, Liverpool. I will not dwell on the mechanical problems of getting her upright – shore holdfasts, tackles, winches, levers, weighting, counter-weighting, and measures to prevent her rolling on her other side at the conclusion – these were the above water aspects which we have already seen in the cases already discussed. On the other hand, the underwater work is well worth noting as an example of the tedious work that is needed for success in harbour as well as in more open water.

Masts, funnels, bridge, deckhouses, davits, lifeboats and cargo winches had to be cut away by divers. The work was carried out under Mersey Docks conditions of almost nil-visibility; this was later improved by a perspex-cone attachment to the divers' helmets giving 230 mm. clear view ahead. Diving difficulties were increased by floating debris and by much diesel and fuel oil leakage which rendered diving dresses useless in under a week. Despite periodical lowering of water in the dock, one third of the painstaking *interior* survey had to be done by divers in similar conditions. Eleven salvage pontoons were secured to the ship at dock bottom to provide buoyancy. Temple-Cox airbolts were shot through the shell plating by divers from inside and compressed air was blown out to loosen the external section of the mud. When upright, 184 scuttles had to be blanked off and three access doors welded. And at the air-water interface, 3,000 tons of mud was discharged by hand.

On 27 June 1954, seventeen months after sinking, the vessel was refloated. It was a copy book operation by the Mersey Docks and Harbour Board staff; master-minded throughout by Captain W.R. Colbeck, RNR, the Marine Surveyor and Water Bailiff.

So far we have seen three refloating cases but none by tidal lifting craft, the operation of which was described in chapter three. One such excellent example was that of the French destroyer *Maillé Brézé*. By curious coincidence I was officer-of-the-watch on the nearby HMS *Furious* when the French destroyer had an internal explosion and sank in 1940. It happened when we were at anchor at the 'Tail o' the Bank' off Greenock and was one of the most dreadful scenes I ever had the misfortune to see. Men burned, men trapped, men screaming with pain, men with arms stuck out of small openings asking for pain-killing injections from our medical officers in boats alongside. It was something I never hope to see the like again.

In 1953, the Ministry of Transport and the Clyde Harbour Authorities became much concerned at the prospect of oil pollution from the wreck and from the possibility of her ammunition becoming dangerously unstable with age. It was decided to remove her.

The wreck was in 36 feet depth at low water and listed about 17 degrees to starboard and it was estimated that after removal of the ammunition and the superstructure etc., a submerged weight of some 3,600 tons would be involved. To cope with this, four lifting craft with a combined lifting capacity of 3,900 tons were employed. These craft were secured, as usual, in pairs, 9-inch circumference wire ropes were worked laboriously underneath the destroyer and the wires hove taut. The main problem here was the exact co-ordination of lift between the four craft, i.e. no single craft must for one moment have more than her share of the weight.

Sunken vessels. Raising. Refloating. Removing

This extremely vital aspect was given the benefit of much advance thought, advance planning, careful calculation of stresses and test trials with scale models. Because of the difference we have already seen between theory and practice, the chief salvage officer remained apprehensive. But it all worked like a charm. *Maillé Brézé* was lifted and shifted into shallower water five times. She was finally beached in 12 feet of water where she was patched, pumped out, refloated and towed away to the breakers yard.

The whole job was acclaimed as a great success and many messages of congratulation were received from the highest authorities. But – as is pretty universally known – every silver lining has a cloud! And our cloud arrived about one year later and rained drops of horrible acid on us. The Treasury declared that the recovery cost of £246,545 was excessive and went on to say that their 'proper function had been largely frustrated by inadequate and misleading information but as the wreck had been lifted, they had no alternative but to sanction the expenditure'. But remember Forsberg's Law – 'a costly success is always better than a costly failure'. Suppose we had given up the job altogether after spending £100,000. The Treasury would have been madder still. You can't win!

In point of fact, the cost was only a notional expenditure, in any case. The Admiralty had the ships, men, material and expertise to carry out the task. The Ministry of Transport wanted the task carried out, and, as a result, handed the Admiralty a hefty cheque. But both parties were government departments and it was merely the government's left hand paying its right hand – with a tiny and insignificant amount ever leaving government custody. It is this sort of accounting system that causes competent and professional officers to go puce in the face and speak forth very uncivil words. No one can adequately pay men for handling dangerous explosives underwater anyway – such willing men are wonderful pearls beyond price.

Instead of heaving a wreck upright, it is – not very frequently – more expedient to cut off much top-hamper and float her upside down. One good example of this was HMS

Breconshire (8,952 tons displacement) a former Glen and Shire liner; she had been the victim of enemy air attack in March 1942 while running the gauntlet on the supply route to Malta. She had struggled into Marsa Xllok at the east end of the island and sank on her beam ends in 20 metres of water.

The wreck contained a considerable quantity of fuel oil and an assortment of cargo which included bombs and ammunition; all of which made operations hazardous and unpleasant, particularly for the divers. She was lying on her damaged port side and the first task was to remove her masts, funnel and superstructure. While this was going on, a scale model was constructed in the Malta Salvage Base; tank tests were carried out to confirm calculations made on the effects of introducing air under pressure into the various available compartments of the ship.

The general idea was to bring the bows off the ground by blowing the forward oil cargo tanks. Then by carefully controlling blowing and counter-flooding to turn the vessel bottom up and refloat her in this condition by the injection of air into the machinery spaces and many after holds. Complete tests having been made on the model several times, always with the desired result, the actual refloating operation was started in June 1950 when the removal of the superstructure and sealing of necessary compartments had been completed.

Under the direction of a former Admiralty Chief Salvage Officer, the late Captain O.T. Harrison, the work proceeded without a hitch and weeks of careful planning were rewarded by the sight of *Breconshire* (albeit a pathetic one) floating on an 'even keel' with about 4 metres of bottom freeboard, i.e. four metres of height showing above the sea surface.

The vessel was finally sold to an Italian firm, towed in a capsized condition to Mesina and beached there. Two years later, she was refloated again on compressed air by thirty-six separate pipelines. She was then towed to Taranto and parbuckled upright in the customary manner – on a sloping beach she was rolled over by the utilisation of eleven sets of shore holdfasts, heavy tackles and pylons on the ship's side assisted by air injection and water ballasting. The idea was tentatively to put the ship back into service but it unfortunately came to naught and she was finally scrapped in 1954.

So ended the saga of the *Breconshire*. A particularly sad saga for me because I had known her in the glorious days of Mediterranean warfare when our lives were in danger every minute of every day at sea and in harbour. I had been First Lieutenant of HMS *Hotspur* and – despite the disparity in the two ships' origins, sizes, roles and armament – *Hotspur* and *Breconshire* had been 'chummy ships'. We officers had eaten and drunk together, gone ashore together, and dodged the bombs together. A tiny bit of my heart was broken up in Italy in 1954.

However, for bottom-up refloating, the feats of Ernest Cox in Scapa Flow must certainly have top place for many years to come. Cox had left school at thirteen, been a draper's errand boy, taken a job for a shilling a week at the local electricity station and eventually rose to chief engineer at Wishaw, Lanarkshire. Quite a successful career in itself for a young man. But at age thirty-eight, he moved into the scrap business and in 1924, aged forty-one, he bought twenty-six destroyers and two battleships – all units of the former German High Seas Fleet scuttled by their crews in Scapa Flow. Very much by trial and error, he became one of the greatest experts in the world at raising sunken ships.

The whole wonderful, fantastic, almost larger-than-life story is told in Gerald Bowman's book, *The Man Who Bought A Navy*. It is quite impossible in a few short paragraphs even to give a precis of the operations. But you will already know, from cases previously discussed, some of the methods and difficulties. Just put twenty-eight cases like those in isolated gale-swept Scapa Flow and you have an idea of the appalling task. Conger eels, whales and seals to hinder diving. Bursting cables, shrieking gales, underwater explosions and fires caused by oxy-acetylene cutting in gas pockets. Fatal accidents and lucky escapes. Anguish and tension on passage as the warships were towed South for breaking up. Four separate attempts had to be made to raise the 28,000 ton *Hindenburg*.

Suffice to say that what started as a business venture for Ernest Cox became his be-all and end-all. He gave everything, received little and thrived on the burden. After eight years unremitting physical toil and mental strain, he was £10,000 out of pocket. But he did not care because his scrap business had prospered and he had enjoyed, far beyond anything money could buy, the demanding exciting years at Scapa. In 1949, he sold up, became semi-retired and lectured for charity on the Scapa Flow salvage until he died in 1959. Men don't often come like that nowadays or, for that matter, at any time. He was human all the way through – tremendous energy and drive, a very bad temper, possessed of a descriptive vocabulary, stubborn as a mule and generous to a fault.

Reluctantly, I must begin to wind up this chapter. With just a few passing moments of admiration for various odd salvage jobs. First, for the sheer ingenuity of the Canadian salvors who lifted the ore-carrier *A.M. Byers* from the bed of the Saint Clair River between Detroit and Lake Huron. This ship was fitted with a heavy duty Goodyear ore conveyor belt; so the salvage crew seized on it, made a giant collision pad from sections of it, joined the sections together with their own belt-fasteners. They backed it with steel netting, secured it firmly over the extensive collision damage and the *A.M. Byers* was safely in dry dock only eighteen days after the collision.

Second issue of admiration for the salvors of the *Reuben* in Mansa Bay, Tanga. This vessel was sunk by shell fire in World War I and had laid on the bottom for over forty years. In 1956, the ship was raised by Italian salvors in seventy days after the repair of twenty-three shell holes. She was actually raised because of the value of her 1,300 tons of scrap metal. But as a bonus, her 1,500 tons of coal cargo was found to be in first class condition and was sold readily.

The third note of admiration is for Lieutenant T.H. Stevens. In 1862, he raised the yacht *America* from the St Johns river bed seventy miles above Jacksonville, Florida. After expert application of every possible improvised seamanlike device, he was about to admit defeat when a backwoodsman came forward and said she had been scuttled with five auger holes.

Whereupon, Lieutenant Stevens (not a salvage officer but an intelligent seaman) improvised some flumes (cofferdams) pumped *America* dry and plugged the holes.

I end with enormous non-admiration for gentlemen who are going to raise the *Titanic, Lusitania* and *Andrea Doria* from the Atlantic seabed with their own 'special' ideas and a shoe-string budget. With general non-admiration for inventors; in every 1,000 ideas there must be one good one but maybe I have only scrutinised 999 so far. Particular non-admiration which has lasted for very many years, for that genius already mentioned who wanted to refrigerate all the water in Port Said wrecks. But my Supreme Non-Admiration Award goes to the would-be salvor who proposed to raise the *Seawise University* (formerly the *Queen Elizabeth*) from the Hong Kong harbour bottom by the injection of ping-pong balls. We do get 'em, don't we?

Chapter 6

Stricken vessels. Ocean towage

The reasons for vessels requiring a tow from open sea to harbour are many. Loss of a propeller or rudder, fracture of shafting between engine and propeller, collision damage having trimmed a vessel so 'stern-up' that the propeller cannot grip the water, fire or explosion in the engine-room, unplanned shortage of fuel, a ship listed over so far as to make employment of personnel in the engine-room too hazardous. And so on and so forth. some time ago, in one particular quarter-year, I noted twelve vessels losing their propellers at sea. Apart from long distance towing, a tug may even earn a salvage award merely by holding a blazing ship across the wind so that flame, smoke and heat blow away to leeward in order to permit firefighting to commence or recommence. And because of the hazards involved, a very worthy award too.

The history and development of ocean towage runs roughly parallel to the overall history and development of steam propulsion afloat. Fascinating subject though that undoubtedly is, it requires a volume (or series of volumes) to itself. Here I can mention only a few isolated random points which have special relevance to the practice of towing.

Of course, as already said, there were many oldtime cases

of sail towing sail. But they were *ad hoc* operations. Anson referred to them in his *Journals*; Nelson referred to them in now-published letters. Closer inshore, other unsophisticated measures were favoured. I quote the Master of the *Ayde* at Bristol on 2 May 1578: 'Little wind. We hired two men and twenty-two oxen to draw our ship into the river mouth'. But despite episodes like those, the advent of steam truly was the beginning of the towage story.

So far as the origin affects us practically, it was the *Charlotte Dundas* in 1801. With a single stern paddle wheel, she towed two barges of 70 tons each for 19 miles in six hours. Not much of an achievement in these advanced times of super-tugs and super-tankers but that was the start of a whole new maritime industry.

The first steam tug to enter naval service was the *Monkey* in 1821 – a paddler of course. Although an American, John Fitch, had fitted to a yawl a screw propeller of sorts back in 1796, it was Swedish John Ericsson who, in 1837, constructed the type of propeller which remained basically the same for more than a century. The new concept, however, did not receive an exactly rapturous reception. In 1837, in a demonstration, the screw tug, *Francis B. Ogden* towed the Lords of the Admiralty in their own barge at 10 knots. Despite that unique experience, their Lordships remained unconvinced and stated that: 'It is based on erroneous principles and full of practical defects'.

Their Lordships maintained that view, at least in part, until April 1875. They then matched the paddle tug *Alecto* against the screw tug *Rattler*. Both were of the same displacement and horsepower and they were connected stern to stern with a towrope. When the starting signal was given, *Rattler* established an early advantage and finished by dragging the struggling *Alecto* astern at 2·8 knots.

In that same year of 1875, the Thames tug *Anglia* (known from her funnel array as 'Three Finger Jack') towed the Allen liner *Syria* 4,135 miles from St. Helena to Southampton – the longest sea tow to that date. But, from that time on, the distance records went up and up. In 1937, I was in the runner-crew of the disabled *Kingswood* when she was towed from Port Pirie to the Tyne by a Smit's tug *Ganges*. I then believed that, as the distance of 13,500 miles was more than half way round the world, it would never be exceeded as a ship tow. How mistaken can one get? In 1957, Smit's *Clyde* and *Ocean* towed the obsolete Argentine cruiser *Moreno* 14,500 miles from Puerto Belgrano to Hikari, Japan. That distance was considerably less in straight line as the albatross flies. But ships cannot always proceed in a straight line – bits of solid land get in the way. Dredgers, floating cranes and floating docks have been towed over even longer distances.

In jumping from 1875 to 1937, I skipped over one of the most unusual tows of all times – that of Cleopatra's Needle now standing on the Thames Embankment near Charing Cross Station. The needle encased in an architect-designed ship-shaped, watertight pontoon was towed from Egypt to the Bay of Biscay by the steamship *Olga*; there it was lost in a storm and presumed sunk. Luckily, it was subsequently found by the SS *Fitzmaurice* and towed into Vigo. The needle was finally collected from Vigo and delivered to London by our old friend 'Three Finger Jack'. The journey from Alexandria commenced 21 September 1877, and finished in the Thames on 20 January 1878 – around 3,300 miles allowing for detours. With the advantage of hindsight and a century of now accumulated experience, it was evidently faulty thinking to let such a touchy 'vessel' be towed through the Atlantic and Bay of Biscay in mid-winter.

But, interesting as all those sea-towing items are, they are in the nature of planned tows. The vessels were specially made seaworthy before commencement, special towing arrangements made, and they were ballasted to get the best draft and trim. Probably steering facilities were provided and, in one case anyway, propellers removed to minimise drag. Such planned tows may have plenty of exciting episodes but emergency salvage tows come in a super-league of their own. In emergency cases the tug may arrive to find the casualty already in a post-accident shambles with the only way to get to safety perhaps through an Indian Ocean monsoon, a China Sea typhoon area or into the central Atlantic in the hurricane season. Emergency tows probably reached peak volume with the work of Admiralty Rescue tugs in World War II. But, even in the days of piping peace, there are still plenty of emergency cases occurring on the high seas.

At this stage, it will be useful to look at connecting-up techniques between two vessels at sea. From an office desk or fireside chair the procedure seems simple. The thought is that the casualty and the tug are in precisely the same wind and sea conditions and passing a tow rope cannot be too difficult. That, by the use of the tug's engines, she moves close to the casualty, hands over towing equipment and then remains neatly stationary while everything is made secure. How wrong! How desperately, awfully, totally wrong!

Just visualise the windage 'silhouette' of various types of vessel and it can immediately be seen that a conventional cargo steamer will lie nearly across the wind as will a cruise liner; that a ship with engine and accommodation (perhaps a bridge too) aft will tend to lie head into wind. A vessel damaged forward may assume a head-down, stern-up posture and will lie head to wind. A vessel damaged aft, may do exactly the reverse. But those statements can never be more than general guidelines because every single case is different. 'Suck it and see' is not a very seamanlike, or a very elegant, precept but is a most effective one in determining the 'angles of rest' of different vessels stopped at sea.

'Stopped' did I say? That's a wrong word for a start. A ship is not like a car, train or wheelbarrow and – unless she is hard aground or in a drydock – is never stopped. Tides and currents move her bodily around even when seemingly immobilised but this is a comparatively unimportant factor when connecting a salvage tow because both tug and casualty are affected very similarly. It's like two people standing on an escalator – moving over the ground but not moving in relation to each other.

Leeway, or drift downwind, is quite a different propositon altogether. The tug is comparatively small, compact and with a fairly deep grip on the water; she will normally lie stern to wind and drift very slowly; on the other hand a cruise liner with shallow draft and immense areas of windage in the superstructure, lying beam to wind, will sail down to leeward pretty fast. Incidentally, don't make mariners laugh by pronouncing leeward as spelt – it's 'loo'ard'!

Take that last case as an example of a tugman's dilemma. If the tug stops downwind of a cruise liner, she will soon be overrun and bumped about by the latter. Such unwanted physical contact can cause serious collision damage to both ships when sea and swell is running high. Now examine the opposite case. If the tug stops in a safe position upwind of the casualty, then the two vessels drift further and further apart – which is no way of helping when the aim is to pass a towrope from one vessel to the other. The answer to the problem is quite evident but none the easier for being so. It is to plan and execute a near-miss situation; when tug and casualty nearly hit each other but do not quite do so. It is simple to say and its degree of difficulty in practice is one big reason why tugmasters are mostly so grey-haired with such anxious expressions.

I remember well a problem almost on those lines when I was in command of a destroyer proceeding from Malta to Gibraltar. The Admiralty (as Ministry of Defence then was) ordered me to tow a large motor fishing vessel from one place to the other as an economy measure. Not a particularly good idea because destroyers are not really designed for towing except in dire emergencies. Despite that unsuitability, all went fairly well until near Alboran Island when the towrope parted in a gale.

A real stinking Mediterranean 'blow' was going on with a swell which was coming nearly 2000 miles from the Israel coast without let or hindrance. The motor fishing vessel was a sturdily built craft designed to take plenty of punishment in the course of duty. On the other hand my destroyer was a thin-skinned, sophisticated, fighting machine costing millions of pounds and she was not built to knock hard against other vessels in a rip-roaring seaway. The days of laying one's ship hard against the enemy and boarding her have long since gone. (A pity – if you will forgive a tiny digression – because I was never able to utilise the cutlass drill learned in my training ship).

So that was my problem. My ship and the tow lying differently, drifting differently, and going up and down on the swell differently – the last like crazy nightmare elevators lurching at each other in passing. Night was coming on and despite the fragility of my expensive ship, it was unthinkable to leave an unlighted, unmanned vessel drifting in the shipping lanes overnight. I sent a radio message to Flag Officer, Gibraltar: 'Am going alongside to re-connect tow. Some damage must ensue'.

In the event I chose to drift down quickly and to lean on the fishing vessel hard. In this way although we were bumping and boring at each other, she could not get free to get a really lethal bash at me. In due course we got reconnected and underway with my hull dented but happily not breached.

Some days later, the Flag Officer entertained me to lunch at Admiralty House halfway up the Rock. As he handed me a glass of sherry, he said; 'I've told the dockyard to be kind over your damage. No doubt every bit of wear and tear since your last refit will be attributed to your barging match at sea the other day'. *Bloody* Admirals! They reckon they know it all. As a matter of fact he was quite right. The incident had been a heaven-sent opportunity to account for quite a few old berthing scars around the ship's side.

Connection of a tow at sea can never be easy. Never. But one or two tips can make it less difficult. First of these is to have a dummy run; that is to lie fairly close to the casualty to assess the individual behaviour of the two ships and to note how they behave vis-à-vis each other. At this stage, or before, the towing gear and securing arrangements must be brought to instant readiness in both ships. There are few things more likely to make a tugmaster dance with rage and burst a blood vessel than to risk his ship, get a line across and then drift apart because something is not ready. The best possible communications must be established between ships – a radio telephone link is best but contact may have to be by loud-hailer, megaphone, morse lamp or radio-telegraphy. Two parallel means of communication are, if possible, an asset; then failure of one system cannot foul up the whole operation.

Such prior plans and actions pay rich dividends. Although the tug will be well accustomed to sea-towing, the casualty has probably never taken part in anything of the sort before. There is another important point too. If an accident (explosion, fire, collision, etc.) has occurred to her, the Master, officers and crew may not be at peak efficiency due to injury, shock or fatigue. So if sea conditions allow, it is good value to send a tug officer and a couple of good men across to advise and help. The tug must also have her best helmsman at the wheel, the best seamen on the forecastle-head with the best heaving lines and the best shot with line-carrying projectiles in close attendance.

All these preliminaries are necessary so that the two ships may remain dangerously close not one instant longer than essential. Both bow and stern of the casualty are weapons potentially lethal for the tug. The bows are sharp and specially strengthened; should they catch the tug on their downward pitch, they are able to inflict vital tearing and slicing damage. The stern (this may have to be the towing point if the bows are damaged) is also strengthened, generally has propeller blades protruding menacingly and, in many ships, has an over hang. The tug will not get mixed up with that lot in a seaway without getting very wounded indeed. A collision at sea – as a naval pundit once observed – can spoil one's entire afternoon!

A line carrying projectile system of some sort is part of the lifesaving apparatus of all well-found merchant ships. Properly aimed and assisted by wind such a system will achieve immensely greater distances than a hand-thrown line or even a line-throwing rifle. Too often, however, it negates that advantage by catching in rigging, masts, funnels or cargo gear and it takes valuable time to get it clear. I have seen strong silent Mates reduced to jibbering jelly when trying to retrieve an incredibly light line fluttering gaily in the wind from a radio aerial 25 metres up aloft.

Some modern tugs (of which more later) are fitted with bow thruster units to give additional manoeuvrability. But others, however, are still in operation with conventional screw propellers only where the motive force and the manoeuvring force is right at the stern. This factor, together with the natural tendency of most tugs to lie stern to wind, makes it preferable to approach the casualty by steaming downwind; this ensures the maximum response when adjusting the 'stopped' position of the tug. The stern-to-wind attitude also unfortunately ensures that all funnel gases sweep over the bridge and make life almost intolerable for bridge personnel. But that discomfort just has to be endured. As the Petty Officer said to the Ordinary Seaman: 'If you don't like it, son, you shouldn't have joined'.

The best method of approach, of inspired juggling with engine-movements, of intuitive response to temporary changes in wind or wave conditions cannot be detailed in a short general book like this. In fact, it is doubtful if any book can do more than point out what to look for. Deepsea towing is one of the last bastions of what aviators call 'flying by the seat of one's pants' and of making things up as one goes along. A successful link-up with another ship is made as a result of a Master having served long years at sea, having passed the necessary professional examinations, having gained specific experience as a deck officer in tugs and, above all, having been frightened so many times that fright no longer registers.

Let's assume that a rocket or gun line has got to the right position in the casualty. It is a slim weak line in itself and something more substantial is required quickly. First then a heaving line or a point line is attached and hauled across. When that, still smallish, line is in hand a medium sized rope is attached inturn and that is hauled across. Only then can the casualty hope to heave over the actual tow-wire from the tug. The last action is to see that the tow-wire is thoroughly and securely made fast. That sounds like teaching Grandma how to suck eggs but even experienced seamen can fail to appreciate the enormous stresses involved.

When two heavy ships have a hankering to proceed in different directions, the performance is impressive. And it gets even more impressive as one approaches the business end of the towing wire, and sees the stretching and hears the groans. Ordinary bitts and bollards have been pulled clean out of the deck. Strong wire 'necklaces' round substantial pieces of superstructure have been used successfully. Or the wire can come through forecastle fairleads, round as many sets of bitts as possible and be backed on to something substantial like the foremast. Good holding points are many and spreading the load equally between them is most advisable.

The finest method of all, though, is to connect the towing wire to the ship's cable. The bow structure, the windlass, the hawsepipes, the normal cable stoppers and the cable itself have all been specially constructed to hold the ship safely 'come Hell or high water'. There is one main snag which sometimes prevents adoption of this ideal solution and that stems from the impossibility of connecting to the ship's cable while the ship's anchor is still attached. And to persuade any shipmaster to disconnect his trusty and well-loved anchor and drop it irrevocably to the bottom of the ocean is difficult. That is the understatement of the year. It sometimes requires the smooth talk of a double-glazing salesman, the repetitive insistence of a nagging wife and the patience of an elephant. Given sufficient time the anchor can be 'hung off' and retained. That pleases everybody except the bos'n who had to do the 'hanging off'.

Let's leave the casualty and take a look at the tugmaster's problems. His end of the wire will be on a drum of a towing winch in most cases – some of the older may have only a spring-loaded towing hook but there are not very many tugs using hooks outside harbours today. Whichever equipment is in use, the strain must be put on gently and that is a considerable task when it is necessary to overcome the inertia of a big stationary ship. There are absolutely no medals for ringing 'Slow Ahead', turning resolutely to the desired course and bringing the towing wire out of its catenary to a state of violin string tautness. Violin strings sometimes snap and so do tow-wires if treated unkindly. There are absolutely no medals for that either.

Instead of using the somewhat coarse graduations on engine-room telegraphs, I have myself preferred to use a telephone to the engine-room and to ask for a few revolutions at a time as the situation develops. When the towing weight is fully on and the catenary of the wire satisfactorily adjusted, one can think again and revert to engine-room telegraphs. In many modern tugs, there is full bridge-control of the engines and that enables the Master to apply the revolutions very precisely himself. That is a wonderful piece of equipment but by the time it arrived, I was, sadly, in command of a desk instead of a ship.

Just as very gradual working up to the passage speed is wise, so is a very gradual alteration of course. Five or ten degrees at a time will get one safely to the final course – with the precious towing wire still intact in one useful piece. 'Softlee, softlee, catchee monkee' is the finest advice in the tow business – advice borrowed from a non-nautical discipline but appropriate in the extreme.

Once under way, things become easier. But the tow-wire catenary must be watched intently. If it comes nearly out of the water there is too much strain; if it comes quite clear you are probably in trouble. Self-rendering towing winches should prevent too much strain coming on but, from jaundiced experience, I have found they do not always live up to their name. Conversely if, in shallow waters of Channel or North Sea, the catenary drags across a wreck on the seabed, the wire might part that way way instead. Remember Bos'n Murphy's Law: 'Anything that can go wrong, will go wrong, usually at the worst possible moment'. Too shallow? Too deep? Just right? Only the tugmaster, whose neck is on the block, can decide.

Almost as important is to guard against chafing the wire and thereby against weak spots developing. It is most important in the disabled ship which is not designed for being towed, and is unavoidably awkward with sharp 'nips' to the wire as it goes over deck-edges and round awkward corners. Packing the bad places with wood, wedges, grease, sacking and rope wrappings, is good but can never give complete protection. In addition the wire must be regularly hove in or paid out a little to 'freshen the nips': so as not to let the same areas of wire take all the chafe all the time.

Chafing risks are increased by the fact that the towed vessel almost never follows obediently behind the tug. She will frequently fly out to one side or the other – tugging sideways at the tow-rope for all the world like a disobedient puppy on a lead. This problem is worst when either the disabled vessel is unmanned or cannot, for some other reason, be steered. But, even with a constantly vigilant helmsman, it is impossible to eliminate sheering altogether. Alterations in speed and (where practicable) alterations of course often produce temporary improvements in the behaviour of the tow.

All in all a successful salvage tow is the outcome of tremendous professional skill allied to extreme and constant vigilance. A half-knot difference in speed, a bit of weather dodging, great navigational expertise (a tug and tow may navigationally equate to a vessel half a mile in length) and a sixth sense about when things will go wrong if not dealt with immediately.

I have expounded on towing techniques in greater detail and at greater length in this book than on other branches of salvage. It is because deepsea towage is mostly 'out of sight, out of mind' and the task greatly underrated by the public and even by some seafarers. When the television newscaster announces that MV *Blank* is in dire distress, but that the tug *Dash* will arrive in an hour's time, the implication is that everything will soon be fine and dandy. It might well be exactly like that too. But not without much blood, sweat, toil and (occasionally) tears of frustration from a small bunch of devoted seamen.

Of course the most expert tugmaster on the high seas

cannot command success; he can only deserve it. In January 1952, for instance, the whole maritime world watched the English Channel approaches in suspense. For days the tug *Turmoil* dragged the disabled American freighter, *Flying Enterprise* nearer and nearer to safety with painful slowness. At the same time *Flying Enterprise* heeled over further and further on her side with equally painful slowness. The brave battle was lost when she finally turned over and gurgled to the bottom no more than a half a day from the safe port of refuge. It was one of those dogged attempts which deserve success but do not get it.

Merchant ships get ever bigger. Likewise tugs get more powerful, faster and more sophisticated every year. Describe the super-tug of today and it is inevitably the poor relation of tomorrow. There are still a number of World War II tugs giving wonderful service and there are also wonderful new tugs coming off the stocks the whole time. The former must ultimately disappear however and the tugmasters difficulties will be just that tinier bit easier.

Weather-routing probably does not fall under the precise definition of ship-handling but it is a practice of great assistance in dodging the worst peaks of bad weather which make for towing's most dangerous and difficult interludes. Modern salvage tugs have a facsimile weather chart receiver which gives considerable advantages over and above relying on even the most accurate forecasts from ashore. With facsimile equipment, the tugmaster is able to study a detailed weather chart constructed with the vast amount of information and expertise available to meteorological offices ashore. Then he is able to study his up-to-date weather absolutely on the spot and to apply that, with skilled interpretation to make a very specific and accurate forecast for the area close around him. Facsimile is a wonderful piece of equipment. One end scans a weather chart ashore and transmits it; the other end receives it and re-draws it in the ship. It is of great use to the tugmaster in selecting a course to keep clear of the highest and most hazardous sea/swell combinations which can cause so much damage to tugs and their tows.

A well found modern tug will be, of course, fitted with excellent radio and radar, plus at least one (and possibly two or more) electronic position finder and a Telex machine. She probably has two towing machines with winch drums which carry 1500 metres of heavy towing wire on each drum – the minimum of physical man-handling is required. She has good workshops and diving facilities and two large cranes for getting salvage machinery and equipment in and out of the ship.

When you read this, the supertug we describe may well not be the newest, biggest, most powerful tug on the seven seas. But for many years she will be an aristocrat of the towing world and well worth knowing about. Again, however, one must bear in mind those ancient World War II tugs which are still almost unbelievably performing extremely satisfactorily in different parts of the world; they may have been modernised, had diesel propulsion substituted for steam and had modern equipment installed. But they are, in the most part, still the same tugs that those wartime shipyards designed so carefully and built so well to last. If the tug of today lasts as long, she will have been a fine investment for her forward-looking owners.

Having written at length about the dedicated professional masters of the towing world, let me add a few words on the successes of those ordinary shipmasters who have 'had a go' – the enthusiastic amateurs of whom the British heart is so fond. A salvage tow with a salvage award may come once only in an average seaman's lifetime or, more probably, never at all. When opportunity knocks, it is wise to open the door and to grab the opportunity as firmly as possible.

Way back on 22 June 1936, the Australian *Mugana* sent a distress call that her propeller shaft was broken and that she was drifting into danger. The Blue Funnel *Nestor* went straight to the spot but found the water already so shallow that she was severely restricted in manoeuvre. *Nestor* tried drifting lines down on floats; she tried to lower a boat; she fired rocket lines and eventually the very last one managed to get on target. A 'messenger' rope was pulled across by the rocket line, a tow rope was pulled across by the 'messenger'. *Nestor* then towed the *Mungana* clear of danger and, finding things going well took her 170 miles on to Adelaide. Over half a century ago but that splendid achievement is still yarned about by greybeards in pubs and clubs.

In 1940, a Blue Funnel vessel was involved again – in fact two of them this time. The *Telemachus* broke down off a lee shore in the East Indies and, together with her cargo, was valued at £677,000. The *Tantalus*, together with *her* cargo, valued at £465,000 went to assist. The financial figures no longer sound impressive by current standards but by 1940 standards the risk was enough to make the hair stand on end, especially as *Tantalus* was quite voluntarily leaving safe waters to enter dangerous ones. In the event, she successfully towed *Telemachus* out of her exposed position, through a mined area and finally through the difficult entrance into Sourabaya harbour. It then transpired that, because both vessels belonged to the same company, there were legal snags about making salvage claims. A compromise was found by awarding the Master £800 and making £1,800 available for division among the crew. All's well that ends well.

In November 1948, the Canadian steamer *Seaboard Star* was also rescued by a ship of her own company, the *Seaboard Ranger*. This particular pick-up was by intention as the weather off the Central American Pacific coast was reasonable and the *Seaboard Star* was able to wait not too uncomfortably for four days while her sister ship steamed to the scene. The latter signalled in advance for the casualty to have her anchor cable disconnected from the anchor and ready for use.

On arrival, the sea was relatively calm and the two sister ships fell into identical 'angles of rest' and drifted at exactly the same rate, i.e. there was no relative movement between ships when 'stopped'. In these particularly favourable circumstances, the rocket line was fired across, followed by a thin rope, followed by a medium-sized rope, followed by the anchor cable. The *Seaboard Ranger*'s own anchor cable was dragged aft to a joining up spot and the two ships' anchor cables were shackled one to the other. This 'all chain cable' connection between the strongest point of both ships secured them as safely as a pair of Siamese twins. The tow then proceeded 1,330 miles to Los Angeles at a splendid average speed of 7·21 knots with about 210 metres of heavy chain cable between them.

The all-cable method is one advocated for many years in *Lloyd's Calendar*. More recently, since the advent of efficient and reliable man-made fibres, the *Calendar* has also favoured the use of thick synthetic ropes run out from the towing ship and shackled to the cable of the towed ship.

Again on 17 September 1960, the all cable method was employed when the *Journalist* went to the *Parkgate* in the Indian Ocean. The connection was pretty meritorious on this occasion because *Parkgate* was high out of the water for lack of cargo and was blowing balloon-like downwind in a fresh monsoon at a rate of about five knots. Nevertheless by good communications and good seamanship in both ships, the whole operation from first rocket to readiness for tow was only fifty minutes. The tow then proceeded for 1,088 miles to Mombasa in eight days and twenty hours – an average speed of 5·1 knots. Not content with that considerable success, *Journalist* went on to tow her charge right into Kilindini harbour through a bending channel only 800 feet wide in places.

As said before, a salvage opportunity may come only once in a lifetime so it must be grabbed with both hands. I remember way back on 9-10 March 1954, the homeward bound steamer *Temple Bar* (with 10,000 tons of manganese ore on board) and the collier *Guildford* colliding in a dense fog off Whitby. About half an hour after the collision, *Temple Bar* had recovered from the shock, found herself in seaworthy condition and made to grab *her* chance – she offered to tow *Guildford* to port. The offer was refused, some five hours elapsed before the first tug arrived and *Guildford* sank seven hours later only half a mile from a suitable beaching ground.

Later, in the Admiralty Court, the owners of *Temple Bar*

claimed the *Guildford*'s sinking was not the direct consequence of collision but was caused by her refusal to take an immediate tow from the ship on the spot. The President of the Court, assisted by Elder Brethren of Trinity House, decided however that it was reasonable and seamanlike for *Guildford* to await the arrival of specialised towing assistance. No benefit therefore ensued for *Temple Bar* but at least she tried.

This chapter could go on for ever but it's time to start coiling up and stowing away. I have spoken about the weather but seem not to have made the point that it is the paramount factor – as it is in all types of salvage. The weather must be humoured, the tow must be coaxed through it and any attempt to defeat weather by use of brute force is almost always doomed to failure.

Despite twenty-five long, hard consecutive years in smallish ships at sea, I am even myself inclined occasionally to underestimate those weather limitations unless I make conscious effort. That is how too long in an office chair decays a seaman. Because of that potential weakness, I have long kept a bit of paper in my right hand top desk drawer to be read about once a month so as to ensure I never forget the hideousness of the cruel sea. It is a tug's *Report of Proceedings* and I will quote only a few short extracts. Conrad could not have written more descriptively.

> All watertight doors, ports, etc. had been closed and everything likely to move about secured. Because of the increasing strength of the wind, the tanker's bows were trying to fly off into the wind, and consequently there was greatly increased strain on the wire . . . our speed was down to 2 knots and we could not keep on course.
>
> The seas by afternoon were mountainous with very low cloud. The ship occasionally rolled so heavily that the ship's bell just forward of the bridge was ringing every few minutes. We took our meals standing up with our bodies jammed against a doorway. Most of the gear we had secured below was strewn about the cabins. The barometer

> was falling about one point an hour and by Sunday afternoon was down to 982 millibars and still falling.
>
> By 1700 the wind was about 60 knots, seas were higher than the tug herself and the ship was rolling 40 degrees most of the time with an occasional roll of 50 to 60 degrees. We knew we were about ten miles from the rocks, visibility was a few yards only and we had not been able to see the tanker for several hours. Another 60 fathoms of towing wire was let out and we eased out a couple of feet every 5 to 10 minutes. By 1900, the wind was a steady 80 to 90 knots with gusts of hurricane force which made the whole ship shudder and vibrate.
>
> As the ship went down in the trough of the waves, the wind whipped the tops of the waves off in a driving, flying sheet of spray and spume. It was also raining hard and a change of pressure as the wind blew in terrific gusts caused an unpleasant clicking in one's ears ... In the dark everyone just held on to anything in order to keep balanced. By 2000 the barometer was down to 960 millibars. We all got very bruised and we were half expecting to beach on the rocks. We had no idea of our speed and no radar. Our only communication with the tanker had broken down as water in the radio room had put out of action every transmitter and receiver.
>
> Some of the rolls we did were terrific and we thought we were going to turn over. There was still tension on the towing wire so we knew the tanker was at the other end but did not know what her heading was ... Water was getting into the engine room faster than they could pump it out ... We were now left with one engine and one generator. The navigation lights shorted and went out ...

Enough is enough. If that Report has put weather into some sort of perspective for those who have not fought for survival at sea, then it has served its purpose well. I personally shall be eternally grateful to Sub-Lieutenant A.H.C. Smith, RNVR, for writing it.

Weather, of course, though of paramount importance is

not the only factor working against success. Not by a far cry. I recall one occasion when one salvor was a major obstacle to another salvor achieving success. That was the memorable occasion when a merchant ship called *Mildred* became disabled in the North Sea. A salvage tug rushed to the spot, got a towline on board and started to tow. Progress was only impeded by one thing. A trawler had also rushed to the spot, got a towline on board at the other end and started to tow in exactly the opposite direction. It was a memorable tug of war while it lasted but the result was really a foregone conclusion. As they say in the boxing fraternity: 'A good big 'un will always beat a good little 'un'.

One other hazard, among a very great many, is the national pride of young emergent nations. In 1961, as just one example, the Djakarta Court confiscated the Dutch tug *Noord Holland* for trespassing in Indonesian waters. The Court, at the same time, sentenced the Master and Mate to two month's imprisonment.

In the words of my old trainingship motto: 'It's a grand life if you don't weaken'.

Chapter 7

Wreck dispersal. Harbour clearance

As can be well imagined, a wreck or pieces of miscellaneous wreckage within harbour precincts can slow down or – in the worst cases – bring to a standstill the operational programme of that harbour. Much the same can be said if the wreckage is in the sea approaches to the port. Even further afield in areas of heavy maritime traffic (for example, the English Channel or North Sea) wrecks can be extremely hazardous to navigation. All such obstructions must be cleared as early as practicable to allow free passage for vessels and craft.

Put very briefly, the rules of the game around the British Isles are these. In harbours or tidal waters under the control of Harbour or Conservancy Authorities, these authorities are responsible for raising, removing or destroying wrecks therein. In naval ports the Queen's Harbour Master is similarly empowered. In open waters the Lighthouse Authorities – Trinity House, Commissioners of Northern Lighthouses and Commissioners of Irish Lights – have parallel responsibilities. In any locality, no authority has jurisdiction over Her Majesty's ships except the Crown. While action is pending, or being taken, the relevant authority is bound to light and mark any wreck which is a danger to navigation.

As with any statement which tries to concentrate years of custom, practice and legal negotiation into one paragraph, those lines of demarcation cannot be regarded as infallible. But they are a very good general guide. Wreck dispersal is an expensive business and there will always be a dragging of feet in borderline cases, with recourse to legal advice and negotiation.

Right now, a very short historical survey will probably be most useful in clarifying the subject overall. In addition to which there is no doubt that wreck dispersal and harbour clearance reached immense proportions in World War II. And the techniques developed then by absolute necessity have stood us in good stead ever since. Technology has moved on and given us a few better tools but the real fundamentals remain the same – notably commonsense and good applied seamanship.

Before World War II, dispersal work was carried out partly by the Authorities' own resources and a good deal by contractors. But sinkage during the war rapidly overtook the capacity of this system and the deficiencies were further aggravated by Government requisition of practically all spare ships and personnel. After much discussion, it was therefore decided that the Navy should assume responsibility for clearance of all war casualties and the Admiralty Wreck Dispersal Department was accordingly set up. With typical British compromise; Captain G. Curteis, MVO, RN, (Rtd) was borrowed from the Elder Brethren of Trinity House, (the largest peacetime wreck dispersal authority) to become Director of Wreck Dispersal.

The infant Department found itself without vessels, men or equipment – and with every other already established maritime service clamouring for exactly those same items. This is no place for detail. Suffice it to say that on 26 February 1941, the SS *Dalriada* was wheedled out of the Director of Sea Transport. She was a passenger/cargo vessel belonging to Clyde and Campbeltown Steamships and was really too large but nothing else was on offer. *Dalriada* was converted, manned in August 1941 and sent to sea to 'learn on the job'.

There was no 'Manual of Wreck Dispersal', practically no existing expertise and certainly no precedent for such intensive effort in arduous and dangerous war conditions. Starting from scratch, the dispersal fleet became one of the vital factors in keeping our ports open despite all-out enemy effort to close them. I quote just one simple early task and then one complex later operation to demonstrate the range and scope of activities undertaken.

One of *Dalriada*'s early jobs was on the 220-ton trawler *Kennymore*. She was lying cross-tide and listing on a sandy bottom with siltation taking place on one side. In a general depth of some 50-60 feet, there was only about 15 feet of water clear above the vessel's funnel-top. First a diver placed a 100 lb Amatol charge in the engineroom; when detonated, it produced much damage but achieved nothing like the required depth increase. Three subsequent explosions were needed before the stipulated clear navigable depth of 34 feet was achieved. Although only eleven working days were taken actually on the job, the North Sea's winter weather, exigencies of war and teething troubles of the new/old ship stretched the operation to over six months. That was lesson number one. More ships required, better ships and better techniques. As a matter of historical interest, a total of 450 lbs Amatol and three depth charges were used.

An incident from towards the end of the war shows that much had been achieved in the intervening years. It also shows that as the resources and expertise grew, so did the problems and size of task. I quote from a report concerning the clearance operations at Marseilles after the liberation forces arrived.

> They found every entrance blocked and the passages between basins choked with blockships sunk fore and aft the fairways. In the Northern entrance of the harbour it would be possible to use dispersal methods because the depth of water would allow of wreckage being left on the bottom but at the Southern end, Joliette and Vieu Port, all the debris of the blockships would have to be slung, lifted

> and taken away as it was cut or blasted adrift. Apart from the blocking of fairways, every quay, shed and crane had been demolished and every harbour berth was choked with sunken craft and heaps of rubble and dockside cranes blown overboard. In addition, seventy-five large ships were lying sunk at all angles.
>
> The most serious aspect of the situation was the intensity of mining which had been carried out. Besides some 6,000 landmines in the area, 400 submarine mines had been laid in the basins, many of them close to the blockships. On several occasions minesweepers operating *outside* the main breakwaters detonated mines *inside* the basins. The urgency of getting supplies into the harbour made it necessary to push on with the diving work without waiting until all the mines had been cleared and the risks were faced by the divers of both nations with courage and tenacity. Two of our diving boats were mined and lost and two British divers are stated to have survived the explosion of a 1,500 lb magnetic mine at a measured distance of 1,100 feet in the basin, in which they were at work underwater.

Even before our invasion of France, over 100 wrecks had been dispersed by the motley flotilla of 26 vessels. Then at Boulogne, ships, some piled on others, were cleared. At Ostend a channel was blasted through the wrecks. Again at Calais, the channel was blocked by 13 vessels but – by rough heroic methods – ships were moving through the area in 14 days. Further inland, 27 bridges blocking the River Rhine had to be cleared later on.

The end of the war left 428 dangerous wrecks in Trinity House waters and those left in Scottish and Irish waters brought the total to nearly 500. Had the Admiralty relinquished responsibility at that juncture, it would have left the Lighthouse Authorities a quite impossible task with the limited resources at their disposal. In addition, a new peacetime clearance to a depth of 45 feet was decided as necessary to classify a wreck as non-dangerous; since less than this depth had frequently been accepted in war, a great many casualties needed fresh treatment. The Admiralty therefore undertook to carry on the work as agents for the Lighthouse Authorities until the task had been reduced to peacetime dimensions.

The post-war drive started off in fine style with the conversion of eighteen modern 'Isles Class' trawlers to replace the old requisitioned ships – and the aim was to disperse every wreck within a period of two to three years. In the event, frequent and painful cuts were imposed on the Navy by successive Governments and the task was not completed until 1958. For the final three years, the naval wreck dispersal flotilla was down to one single ship; it was assisted very ably indeed by one Admiralty civilian-manned vessel.

A simple question may now arise in the minds of thoughtful readers. If all wrecks were cleared in 1958, why are we talking about it today?

The answer to that comes in three-fold form. First, discovery of 'new' wrecks has gone on over the years. Even today, after many subsequent years of hydrographic survey work, and helped by the computerised wreck record held by the Hydrographer of the Navy. I do not believe that 'Droggie' would put hand on heart and swear that he knew the lot. Second, if you read – as I gloomily do – the court reports of formal investigations, you will find quite a number of vessels descending into Davy Jones's Locker every year. Third, with the tremendously increased tonnage and size of ships in recent years, the 'original 'safe' clearance of 45 feet has become decidedly unsafe. Some of the larger vessels now draw 30 metres and ones using the Channel and North Sea port approaches, although smaller, require 20 metres and upwards. Since some are navigating in areas with little more than a metre under their bottoms it is dangerous – or at least extremely expensive – to have any wreckage at all standing proud of the seabed.

The only certain way of restoring the seabed to its original 'clean' condition is to raise the wrecked vessel (as described in the earlier chapter on refloating sunken vessels) and to take her away from the area altogether. A method almost as certain is to use underwater cutting tools and explosives to divide the vessel into pieces of a suitable weight to be lifted and carried away. But although those are probably far the best methods for close inshore, estuarial and in-harbour wrecks, they are almost always unsuitable further offshore. In those exposed positions, weather can hinder progress and so destroy work already done as to make refloating or lifting impracticable if not actually impossible. In such cases, 'blowing down' is the only reasonable method.

One of the first essentials to any successful wreck dispersal task is to get the appropriate maritime authority to publish a 'Notice to Mariners'. That ensures the careful avoidance of the area by all the most efficient ships. The second essential is to be certain to have the wreck area as brightly lighted and as clearly marked with large wreck-mark buoys as is possible. That ensures avoidance of the area (albeit worryingly late) by those who have not received the latest Notices to Mariners but who do keep a good lookout. The third essential is to keep a thoroughly good look out oneself and to have, at instant readiness, searchlighhts, sirens, loudhailers and pyrotechnics to blast off those incompetent vessels who will blunder through almost anything.

If that sounds over-pessimistic, it is not. On 13 January 1971, *Texaco Caribbean* sank off the Varne Bank after a collision. The following night, the *Brandenburg* hit the first wreck and sank despite the presence of a Trinity House vessel exhibiting wreck warning lights and a radio warning of the danger being given. Because of this double disaster, the danger was marked by a wreck-warning lightvessel and five lighted buoys. But on 27 February the *Niki* sailed into the area and became wreck number three. That was an out of the ordinary case but there are few offshore jobs done without several near misses and frights.

As one example, I remember a passenger steamer passing one of our dispersal teams at high speed in the Thames Estuary. In the shallow water there, the wash from the ship had an enormous effect on the diving boats; they started to heave, pitch, roll and surge with unexpected and dangerous strains coming on the divers' vital and easily fractured air lines. Equally bad, although down below and unseen, the divers were brutally pushed and jostled against the jagged wreckage they were working on. Altogether and alarming episode developing out of sheer unadulterated thoughtlessness. The particular offender was one that did *not* get away; the Master was brought to Court and fined.

In 'blowing down' a ship a big explosive blast (albeit possibly a widespread application of small charges fired simultaneously) is more effective than a succession of smaller blasts. That is the general rule but there are occasions when a single charge placed very accurately will neatly remove a 'high spot' without blowing up another one in its place. Each case must be judged strictly on its own merits; in this sort of work, more haste very truly makes less speed. Days spent in careful preliminary survey are usually more than regained in the long run. The scale of explosive action must, of course, also be adjusted according to the proximity of the wreck to jetties, docks, lighthouses and bathing beaches.

Explosives are either placed in position by divers or 'swept in' on wires. Each method has advantages but the methods are complementary and not in competition. Explosives can be 'swept in' when divers cannot operate because of rough weather, bad underwater visibility, fast tidal streams or low temperature. On other occasions, a diver can accurately locate the best place to apply the explosive to best advantage.

The easiest and most usual method of dispersal is to dig a 'grave' with explosives. (The Navy mainly used depth charges). It is usually found that erosion has taken place on one side of the wreck and the object of using explosives is to expedite the erosion and to shift the wreck down into the trench which is left behind. During the war, the Royal Naval Scientific Service produced data relating to the size of explosive, bubble and resultant cratering in different depths of water. From that data, a ready-use scale was evolved to show how an area of sea bottom could be given maximum treatment with minimum expenditure of material. As one example of efficacy, a 6,000 ton ship with 8,000 tons of cargo was lowered 18 feet by only one set of depth charges. It was not claimed as any sort of record so presumably it was good without being unique.

On hard bottoms, grave digging is unfortunately unworkable. It is then an absolute necessity to place internal charges in exactly the right places. Some of the ship's fittings, especially boilers in steamships, seem almost indestructible. It is occasionally better – although very much longer – to disperse methodically from the top down; then to lift and take away the most obstinate sections.

One thing only is certain in wreck clearance. It is that 'one cannot please all the people all the time'. When the war-time and anti-aircraft towers in the Thames and Mersey approaches were demolished, various seafarers protested against removal of prominent seamarks. If only one single explosive charge is used miles from the nearest habitation, someone will complain of broken windows. If, for the safety of navigation, a wreck is blown to pieces on the bottom, it will be merely a day or so before local fishermen complain of tearing their nets. If, on the other hand, the wreck is dismantled and removed piece by piece until nothing is left on the bottom, the wreck officer will be vigorously assailed for incurring unnecessary expense. This game is almost impossible to win!

On completion of the task, it is essential to check the clear depth obtained over the top of the wreck. Sonar sweeps – even with today's high quality instrumentation – are not infallible; it is impossible to be quite sure that thin but substantial obstructions (say a derrick, stump of mast or boat's davit) will show up on the echo trace. The only acceptable thing is an old-fashioned physical check by drifting over the wreck with a bar-sweep hung over the side and lowered to ever deeper depths until it finally fouls the highest peak of wreckage remaining – a bar-sweep is literally an iron bar suspended by a measured line at each end and (to avoid any bending effect) in the middle as well.

A further check sweep is often carried out by lowering

measured chains over the ship's side and connecting the bottom ends with spurnyarn which will break whenever anything substantial is fouled. If underwater visibility is good a diver may also take a measuring line to the highest peak of wreckage and then measure accurately upwards to the surface.

The 'raw' clearance depth is, of course, no use to navigators. By use of tide tables and calculations, the depth over a wreck must be computed so as to be expressed relative to chart datum, i.e. to that particular low-tide level used on the charts of the district.

Wrecks can be cussed in character and inexplicable in behaviour I recall a former colleague of mine, Captain W.R. Fell being stumped and on the verge of being somewhat professionally discredited by one. He worked one Autumn on a wreck in Scottish waters, painstakingly obtained the necessary clearance to chart datum and reported accordingly. During the next Spring the Hydrographer sent one of his specialist survey vessels for final confirmation before printing the information on charts. The survey vessel reported one foot less clearance than Captain Fell. Black mark!

Captain Fell took his ship back, did some extra work, went through the whole laborious routine again and made his report. The survey vessel attended in due course and once more found it one foot short of requirement. Blacker mark still! On the third occasion, an irritated, frustrated, Captain Fell carried out an intensive and thorough investigation. Mystery solved! One end of the wreck was on a soft bottom, the other end on hard. Each gale – and there are plenty of those on the West Scottish seaboard – scoured the soft bottom away. Then, as that end of the wreck sank, the other end cocked up towards the surface. My relieved colleague cut five or six more feet of superstructure off for good measure – and hang the expense!

That see-saw type of movement is not the only one to be encountered. Ships can be visible on a sandbank, seemingly for ever, but then disappear entirely into the sand – the famous frigate *Lutine* is just such a case. Then again at the beginning of one season, one of our ships discovered a wreck half a mile distant from where a colleague had reported it at the end of the previous season. ('Old Charlie must have had a hangover when he "fixed" this one', the captain signalled to base). A small trawler wreck was not so spectacular but being close inshore was very reliably pin-pointed; she moved 130 metres in seventeen years.

Those movements are not so remarkable when one learns that large and conspicuous sandbanks do not themselves remain stationary. In 1958 for instance the Hydrographer's survey showed that the southern tip of the Goodwin Sands had moved half a mile in eleven years. Finally there have been cases of known wrecks, innocuous for many years, suddenly becoming dangerous: experts can only conjecture that bottom scouring by strong tides may have rolled upright such vessels formerly laying on their beam ends.

So much could be written on wreck clearance and it would all be of absorbing interest. The twice-over clearance of the Suez Canal after two Arab-Israeli conflicts. The frustrating story of Dover Harbour where, between the wars, enormous efforts were put into cutting up and removing a World War I blockship from the western entrance; then at the outset of World War II, one of the very first acts of national defence was to sink a blockship in the western entrance! There was the dispersal; of the *Royal George* (of 'Toll for the brave' fame) which to the great shame of the Navy had to be cleared from the Spithead anchorage in 1839-40 by a soldier – Colonel C. Pasley of the Royal Sappers and Miners. In modern times, the clearing up from Korean, Vietnamese, Indian sub-continent and Arabian (Persian) Gulf conflicts has been as heroic as it had been efficient. The Port of London Authority could tell a good story too; its house journal one year said that the PLA had retrieved innumerable items from the river ranging from a tug to a perambulator.

Rock clearance is second cousin twice removed from wreck clearance. In 1958 I envied Canadian engineers their task of blowing off the twin peaks of Ripple Rock, British Columbia'a worst navigational hazard. That task took 1,400 tons of nitromex explosive and (at 1958 prices) cost £1·1 million. What a job! If I ever spent more than a couple of thousand, *my* bosses were exceedingly grumpy.

Wreck dispersal and harbour clearance usually lack the glamour of real salvage work. But it is good solid, satisfying, seamanlike work and there is very real sense of achievement at the end. I had the good fortune to be the Admiralty Assistant Director in charge of wreck clearance work for the last three years of the Navy's 'sovereignty' over the business and I found it one of the most interesting periods of my life. There was one solitary reason only for being glad to hand the job back to the peacetime authorities – and that was because of the wreck of a small trawler in the middle of some valuable oyster beds.

The oyster bed proprietors were extremely cross with the Admiralty. They held us liable (they said) for loss of productivity of their beds. They wanted the wreck removed forthwith – if not before. Further, they wanted it removed entirely without noise because they said (they did really!) that noise would upset the valuable oysters. That was a ticklish situation indeed for yours truly and it was at this very juncture that we returned the coastal wreck dispersal task to the Lighthouse Authorities.

For long afterwards I wondered if Trinity House had managed to solve that problem and how. Nowadays, it is rather shellfish of me but I don't really care any more.

Chapter 8

Recovery of Submarines

In 1578, William Bourne, a then eminent British mathematician and author, designed an enclosed boat which could be rowed under water; its waterproof leather covering was to be expanded and contracted manually to alter the internal volume and thereby the buoyancy. A well-conceived principle but it was not until 1620 that Dutchman, Cornelius van Drebel, put the design to practical use. He built three submersible boats in succession and was successful in operating them on trials down to 15 feet (5 metres) below the surface; propulsion was by means of oars through tight-fitting apertures in the leather covering. Endurance was governed by the amount of work the crew could perform on a limited amount of oxygen. The history of submarines from then on is a fascinating one but this book must resolutely foresake it and drive on to the salvage and recovery aspects.

So we progress quickly to Mr. Day, an inventive ships' carpenter, who produced a submersible boat in 1773. His somewhat elementary concept was to attach rocks to the keel until the enclosed boat sank, then when the rocks were released by remote control from within, the boat would recover sufficient buoyancy to return to the surface. On a 30-feet (10 metres) dive the trial was successful but on a later, and much deeper, dive the craft went to the bottom and failed to resurface. Possibly there were earlier sinkings but that is the first incident I have seen recorded.

From time onwards, submarines – although getting ever larger, more seaworthy and more sophisticated – have continued to sink with heart-sickening regularity. One terrible and tragic loss in post-war years was that of the USS *Thresher* on 10 April 1963, with 129 naval and civilian personnel. Quite apart from many hundreds of wartime losses, there have been over 100 submarine peacetime losses from sixteen different countries in this twentieth century. That sounds a great number but boiled down to bare statistics is only one per year between all the navies of the whole maritime world.

It must be first emphasised that when a submarine sinks, saving of life is the immediate and paramount commitment. Every effort, every thought and every resource is concentrated on retrieving the people inside. Only when this operation – which could quite well last several months with modern long-endurance submarines – has been concluded, do the salvage and recovery teams move on to the stage. In earlier days, with smaller submarines and shallower operating depths, lifesaving could often be effected by lifting the submarine to the surface with her personnel still inside. Theoretically that is still practicable in certain cases. But with the tremendous increase in possible escape deaths, allied to the tremendously increased weights of submarines, the possibility gets even more and more remote. However one such success with a small submersible is described later.

Naturally before either lifesaving or recovery can be started, the exact position of sinking must be found. Sometimes that position is self-evident. For instance when HMS *Artemis* gurgled embarrassingly to the bottom at her berth in Gosport there was no location problem at all! When HMS *Truculent* was rammed in the Thames Estuary, the position of collision was fairly accurately known; it was then only a case of calculating the tidal stream, estimating the sinking time and marshalling a couple of sonar-fitted vessels to pinpoint the casualty quickly. In worst cases, where the submarine is by herself and fails to resurface, the search is inevitably protracted; HMS *Affray* sank in the English Channel on 17 April 1951, and despite the most intensive efforts was not found until fifty-seven days had elapsed. The search for USS *Scorpion* in very deep Atlantic water was not successful until the 148th day despite the use of every conceivable modern device.

The operations following the loss of USS *Thresher* are specially valuable to study at this juncture. They give an idea of reasonably modern techniques and also of age-old difficulties inseparable from work in the sea. They further go to show that even if full recovery if not possible, photographs, on-site inspections by submersibles and recovery of small items can do much to establish the cause of accident. But even more important, they can prevent another similar one from happening and all this meticulous detective work naturally helps to maintain the high morale of crews of vessels in a similar class.

The *Thresher* was a nuclear propulsion submarine, had originally commissioned for service on 3 August 1961, and was, on 10 April 1963, carrying out test dives after a dockyard refit. Her details, (according to the contemporary *Jane's Fighting Ships*) included a submerged displacement of 4,300 tons, length 84·9 metres, breadth 9·7 metres, draught 7·7 metres and a submerged speed of thirty knots. This splendid vessel sank in just seconds after transmitting a garbled radio-telephone message to USS *Skylark* which was in close attendance on the surface.

Although the pre-accident position was known fairly accurately by the *Skylark*, the submarine's high speed and extreme manoeuvrability during the underwater trials meant that her ultimate position on the bottom could be estimated no better than approximately. So round the best estimated (datum) position, an area of ten miles by ten miles was designated as the search area. First efforts were directed to obtaining good clues in order to be able as early as possible to concentrate and focus the search into smaller compass. Oil samples were picked up from the sea, as were pieces of plastic, pieces of cork and rubber gloves.

The oceanographers then calculated backwards the possible drift of these objects at various depths between time of accident and time of pick-up. These results refined the search area a little more but not very much. For one thing the best navigational aids of 1963 could, in that vicinity, guarantee no greater accuracy than 300 metres – and when looking for an object 85 metres long in water 2,800 metres deep, considerably more accuracy than that is required for quick results.

Listening devices were used, as were active sonar sets, precision echo sounders, magnetometers and Geiger counters. But eventually the best indicators were many pictures of the seabed by deep underwater cameras. The uninitiated might wonder then why the whole area was not systematically searched by means of underwater photography. And the plain answer is that it is quite impracticable, for several reasons, to ensure 100 per cent coverage of an area.

The fact is that if an underwater camera is working in water of sufficient clarity to allow each bottom picture to be 10 metres by 10 metres in size then these are good conditions. With a picture of such size – and provided there were no other problems to contend with – a ship would have to steam some 2,000 miles to photograph the whole area. The ship's speed could be hardly more than a drifting speed or the camera, instead of being under the navigating platform vertically, would fly out like a kite. Add to this kite-effect the different sets of tidal streams all the way from surface to bottom. It can be seen that even if the searching ship could navigate to an accuracy of 10 metres (which even with today's ultra-modern equipment, capable of fixing a position to within one metre, it cannot) there is no means of getting the camera to 'navigate' with equal accuracy down at seabed level. It can be seen by elementary trigonometry that the bottom end of a 2,800 metres wire hanging only one degree from vertical will be transversely offset by 50 metres from the top, or surface, end. Finally, one drawback culled from the harsh and sad personal experience – no electric or electronic equipment functions 100 per cent efficiently for ever in harsh shipboard conditions. At which juncture I instantly lose beautiful friendships with every manufacturer in the electronics industry. But it is true.

Way back in 1558, Queen Mary I of England declared that when she died, the word 'Calais' would be found written on her heart. When I follow her, the words written on my own heart will be: 'The radar (or the echo-sounder or the gyro-compass or the radio-telephone or . . .) has gone on the blink, Sir'. It *always* goes 'on the blink' at the precise moment one needs it most!

Notwithstanding the impossibility of 'searching' by camera, the pictures of debris on the bottom did narrow the search area still further; at that stage the bathyscape *Trieste* was brought into the act to carry out further camera indentification together with the application at the sea bottom of that most infallible of optical devices – the Human Eyeball Mark I. *Trieste*, having once descended to the bottom of the deepest known bit of sea in the world (11,000 metres in the in the Marianas Trench) was more than capable of operation in the *Thresher* vicinity. But she, in turn, had operating difficulties – a search width of about 30 metres, a maximum speed of 2 knots, a mere four-hour endurance on the bottom and a need for considerable maintenance between dives.

In the end, after infinite patience and devoted work by surface and sub-surface personnel, *Trieste* positively indentified and photographed parts of the submarine; she also brought up in her manipulator-claw a piece of twisted copper pipe marked '593 boat' the absolutely infallible shipyard proof of identity.

This, as was said earlier, was not a 'true' operation of salvage or recovery. But it was the most efficient possible substitute. After examination of the flotsam, the single recovered piece of metal, the photographs, the communications logs and the many witnesses, the Court of Inquiry was able to come to a reasonable conclusion. Amongst many generalities, it specifically reported the likelihood that a seawater pipe broke and admitted a spraying, misting, destructive jet of water under enormous pressure.

The Report recommended development of new inspection techniques for material under the slightest suspicion. The proof of the pudding – as the popular idiom has it – is in the eating. The United States Navy did admittedly lose the *Scorpion* in 1968 but almost certainly that was the result of an unhappy armament accident. Apart from that, the United States for many years thereafter had a wonderfully accident-free period; especially meritorious when the great number of submarines she operates is taken into account. The information recovered so painstakingly from the *Thresher* sinking was greatly, even, if not wholly, responsible for that subsequent good record.

The stories of *Thresher* and *Scorpion* make grim reading. They sank in such deep water as to make it operationally daunting even to think of complete salvage – and even if operationally possible the requisite financial outlay would be astronomical. Specialised ships men and equipment would be required, possibly for years, while waiting for and utilising tiny periods of suitable weather conditions. Imagine the weight, in themselves alone, of the great lengths of heavy lifting wires and the gigantic winches required to operate the lifts. Imagine the phenomenal pressure needed to inject air and eject water when the ambient water pressure at, say, 3,000 metres is 300 atmospheres. Imagine also the great weight and susceptibility to fracture, damage or fouling of armoured air hoses hanging all that way down from the salvage vessels. Imagine the huge surging oceanic swell.

Despite all the practical difficulties, however, it is the financial ones which are the most formidable. It requires only a resolute national leader, like the wartime Churchill, to write on a sheet of paper: 'Pray salvage that submarine. Time and expense are of no consequence'. In which case there would be virtually nothing absolutely unrecoverable. 'The impossible we do immediately; miracles take a little longer'.

Recovery of Submarines

One pointer towards the operational practicability of very deep ship-salvage was given by the 'mining vessel' *Glomar Explorer*. In a very remarkable 1974 operation, she recovered one-third of a Soviet submarine from 5-6,000 metres of water north-west of Hawaii. With just a bit more luck, it could have been a complete success. Yet no salvage officer in his right mind would previously have given the project a snowball-in-Hell chance of coming good. Few details are available because it was primarily an important intelligence-gathering venture. But for sometimes stated cost of 130 million dollars, it proves my point that, for a lot of money, a lot of previous unthinkable recovery work can be undertaken.

At this stage, it will be useful to make the point that the methods of refloating a submarine are exactly the same as those already described for raising other sunken vessels. That is to say the vessel must be:

(*a*) bodily lifted by tidal or winch power *or*
(*b*) given back some buoyance by injecting air *or* foam and thereby expelling water *or*
(*c*) given artificial buoyancy by attaching external apparatus, for example camels or pontoons *or*
(*d*) refloated by a combination of the above measures.

As in every branch of salvage, each submarine salvage case is a 'one-off' as regards detail. Depth, weight, amount of damage, climatic conditions, tidal *regime*, water temperatures, whether in or out of ships' traffic lanes – and so on and so forth. So no one can pontificate on the 'right' way to carry out a submarine salvage. The one and only approach is to know the fundamentals, to study the vessel's drawings (or better still to study a sister-ship), to get the best possible information on tides and weather, to carry out an on-site survey – and then to plan and think the whole task right through. And, even then, to be absolutely prepared at *every* stage of the operation to re-plan and re-think the way ahead.

I want for that reason of knowing the fundamentals to speak next of the salvage of the British submarine *Truculent*. Admittedly now an elderly case, it is most useful as a study of a medium-size submarine in medium-difficult conditions and it has the additional merit of being well documented. Any person who can understand the *Truculent* operation can (with interpolation, extrapolation and imagination) understand all other cases of submarine salvage. There are, of course, many easier cases like HMS *Sidon, HMS Artemis*, and USS *Guitarre* which have had their accidents within shallow waters of naval ports and bases. But, despite the passage of years, the *Truculent* salvage operation remains one of the most difficult to be completed in relatively open water.

To set the scene briefly, HMS *Truculent* was in collision with the Swedish SS *Divina* during the evening of 12 January 1950, in the outer Thames Estuary. From a crew of 6 officers and 52 ratings augmented by 18 dockyard officials, only 15 persons survived. *Truculent* was of 1,325 tons surface displacement and finally sank with a list of 15 degrees in 22 metres of depth of water on a sand and clay bottom. She lay almost athwart the strong ebb and flow tidal streams.

Necessarily the first planning decision came with the need to choose the general method of salvage. As is generally the case, a compromise was reached. First to lighten the submarine by injection of air until the deadweight was reduced to 800 tons; then to pick up that residual weight by winch-powered hoisting. Injection of compressed air was made into compartments carefully selected so as not to effect longitudinal or transverse stability. Next, messenger wires were see-sawed under the slightly cocked up stern of *Truculent* and worked into the calculated lifting positions. When these were in place, the messenger wires were used to haul the heavy lifting wires under the submarine's bottom where each would, in due course, bear an equal part of the 800 ton weight.

The then-Admiralty-controlled, powered gantry-type lifting crane craft *Energie* and *Ausdauer* were towed to the scene and secured to heavy moorings designed to keep them firmly in place in both the ebb and flood tidal streams. Each craft was capable of lifting 600 tons by gantry and winch; the resultant combined lifting ability for 1,200 tons was therefore comfortably above the estimated weight of the submarine even allowing for infiltration of silt into the hull and any bottom suction to be overcome. Two lifting strops were passed into *Energie* and two into *Ausdauer*. After the necessary rigging connection and adjustment, each craft took her share of the weight and hove *Truculent* clear of the bottom.

Tugs then manoeuvred the whole floating complex 4½ miles to a suitable beaching place, the submarine was taken over at the top of high water, and then lowered into a new and much shallower bottom berth. After an interval here for easier and more detailed overall survey. *Truculent* was again lifted in the same manner and transported to a still shallower position and put on the bottom at high water. When the tide receded, only 3 metres depth of water remained and it was possible to pump out and patch up the hull before the final tow to Sheerness of 23 March for drydocking.

Those were the bare bones of the operation. Things did not always go consecutively as written; the various parts were processed as men, ships and equipment became available. Events planned to take place later were sometimes leapfrogged into earlier places – a state of affairs which is a rule rather than an exception in salvage work. The job was a seaman-planner's delight – divers, ships, anchors, moorings, pumps, compressors, wires and many other items to be ordered, collected, transported and jig-sawed into the grand overall design.

Written in a few paragraphs, the whole project seems ridiculously easy. But nothing could be further from the truth. The work was carried out in the worst winter weather and had to be temporarily abandoned several times on account of severe gale conditions. The water was so cold that all the divers' fingers were turned to thumbs. The placing of heavy moorings in exactly correct positions was an operation almost worthy of a chronicle to itself. The enormously stiff, cumbersome heavy lifting wires and shackles needed equally enormous skill, strong-arm handling and nerve to get into position. The lifting strops slipped when *Truculent*'s weight was first taken and *Ausdauer* found herself carrying a dangerously large proportion of the 800 tons. No one can ever give sufficient praise to the salvage officer, Masters, officers and crews who were (and still always are) dedicated enough to strive for near perfection in conditions when most sensible men would be scuttling for shelter.

Although I have written at some length about *Truculent* because of personal knowledge, I do not want even to seem unduly to be waving a British flag. I suspect that the finest-ever, open-sea operations were right back in 1939 by the United States Navy. The USS *Squalus* had sunk some nine miles off the New Hampshire coast in about 80 metres depth on 23 May. In a brilliant lifesaving phase, thirty-three men were recovered from the submarine by rescue chamber. The accident occurred at 08.40; at 10.15 next day a diver shackled on the downhaul line for the chamber and at 11.15 the first seven men commenced their half-hour voyage to the surface and safety.

The rescue chamber had been devised by Commander Allan R. McCann. It was a pear-shaped container with a thick rubber gasket on the steel bottom rim and the whole contrivance was hauled down by the rescuers to surround an escape hatch; a watertight seal was then effected by expelling water from the bottom compartment of the chamber which thus induced a strong downward sea pressure to compress the gasket. It was then possible to open hatches and transfer personnel from the submarine to the chamber more or less in the dry. Hatches were shut again, the seal eased off and the chamber hoisted up to the surface.

The McCann chamber was the finest submarine escape procedure for many years until finally replaced by the 'free escape and float to the surface' method for depths of less than about 160 metres. And by the Deep Sea Rescue Vehicle (DSRV) for greater depths. A technical description of the latter is out of place here but it is essentially a submersible with the capability of navigating, locating and locking on to an escape hatch without divers' assistance. I apologise for this two-paragraph digression but the subjects of personnel rescue and submarine salvage constantly jostle each other cheek by jowl and it is impossible (and indeed improper) not to give the former brief mention. But now back to the *Squalus* salvage!

In essence, recovery of the 1450 dead-weight tons submarine was effected by another combination of the basic ingredients already stated. Six chain slings were swept in underneath the *Squalus* and lifting was done by a clever attachment of salvage pontoons at different depths on the lifting wires which ran vertically up from the chain slings to the surface; the pontoons in total gave 750 tons of buoyancy externally. The remainder of the buoyancy was obtained by blowing a requisite number of compartments dry with compressed air and ensuring that venting arrangements would permit excess air pressure to escape during the ascent.

The first attempt was abortive in that *Squalus* blew to the surface too fast, 'jumped' in the slings, broke them and returned to the bottom faster than she had come up. It was a matter of going back to square one and starting again – a state of affairs that every experienced salvage officer regards sadly as almost normal. It took exactly one month to get ready again and the second attempt was successful. She was towed to the Navy Yard at Portsmouth with two stops for rest and adjustment of load on the way. She arrived back in the Yard 113 days after leaving for her ill-fated dive.

Despite the importance of the big submarines, one of the finest and most breath-taking operations in modern times was the recovery of the submersible *Pisces III.* The craft was finally hauled to the surface with only a few minutes ration of oxygen left for the two man crew. The tiny sub was 5·82 metres in length, 3 metres in breadth, 3·64 metres in height and weighed 9,700 kilograms when submerged. Thus by weight and size, salvage operations were right back to earliest days. But the depth of 500 metres, the weather-exposed North Atlantic position and the desperately short time available for lifesaving made the proceedings absolutely unique and historic.

Pisces III had been working on a Post Office telephone

cable, burying it in a trench to prevent it being damaged by vessels trawling in the vicinity. At 09.22 on 29 August 1973, she had returned to the surface and was shortly to be hoisted aboard the support ship *Vickers Voyager*; a towline however fouled the after-sphere hatch, water was admitted and the craft sank with both crew members, Roger Mallinson and Roger Chapman, still aboard. Life support was initially assessed as sufficient for two days. But in the event, by careful rationing, by calmness and by avoiding movement, the oxygen supply lasted seventy-five hours and fifty-five minutes.

The first twenty-four hours were spent by Vickers in mobilising rescue forces and arranging for collection of all the necessary equipment. *Pisces II* was ordered in from a job in the North Sea. *Pisces V* was requested urgently from Canada. Arrangements were made for air and ship transport to the site, including crane berths and linking road transport. The American unmanned CURV (Controlled Underwater Recovery Vehicle) was offered by its owners and gratefully accepted. Similar transport plans were made for it too. The Royal Fleet Auxiliary *Sir Tristam* arrived on the scene and took aboard the Operational Controller with portable underwater communications system; this move was to enable *Vickers Voyager* to leave the accident position in order to go to Cork and embark the rescue submersibles.

A more suitable vessel, the naval survey ship HMS *Hecate* arrived eleven-and-a-half hours later and took over the communications team from *Sir Tristam*. And the United States salvage vessel *Aeolus* also arrived on station. The essential preliminaries were going well. At 01.00 on 31 August *Vickers Voyager* arrived back in the area with *Pisces II* and *Pisces V* embarked.

One short hour after arrival of *Voyager*, the *Pisces II* was launched and she descended with a large diameter polypropylene rope; its buoyancy unfortunately proved too great and the submersible's manipulator arm was bent as a result. *Pisces II* was therefore obliged to return to the support ship for repairs. Next, *Pisces V* went down with a thinner buoyant rope and she searched for six-and-a-half frustrating hours before locating the casualty. She attached the rope with a snaphook to the correct lifting point but it unfortunately rolled out 10 minutes later. By quick thinking and manoeuvring, *Pisces V* managed to catch the loose end and to hook it on the casualty's propeller guard. This position was in no respect suitable for a bodily lift but it did provide the first positive physical contract with the surface.

Pisces V then tried valiantly to do more work in order to make the improvised connection into an effective lifting arrangement. But with failing batteries, she ultimately had to content herself with resting on the bottom alongside the casualty – to give moral support and to assist in homing the next submersible quickly into position.

The repaired *Pisces II* now started down to take over the task but the luckless craft had to abandon her mission almost immediately because of a leak in her own sphere. Eight hours later, she was again ready, descended with a rope and was successful in connecting it with a spring toggle inside the hatch of the bottomed submersible – a firm lifting connection at long last. It was however considered safer to use the now-arrived CURV to try placing a second line and a second toggle similarly. After the previous setbacks and frustrations this phase went splendidly. Remote controlled from the surface, CURV located and made another firm connection in under an hour.

Thus two efficient lifting ropes were connected, together with one 'lazy' line on the propeller guard as a possible bonus. At this stage it was decided that the American cable ship *John Cabot* was ideally suited to undertake the long hard careful lift and she accordingly took all lines on board and assumed full responsibility.

After another heart-chilling period, in which CURV's own control lines tangled, with the lifting ropes, the casualty was lifted to within some 30 metres of the surface. At that level, divers – despite heavy plunging of the *John Cabot* and agitated water conditions – managed to pass another nylon rope through a large shackle on the proper well-tested lifting point. All ropes were now hove in gently together. Finally it was possible to attach a really suitable size rope to the lifting point and the salvage was virtually over.

In a paper in the September 1970 *Underwater Science and Technology Journal* I wrote:

> So far submersibles have operated with success. Only one, *Alvin*, has been lost – and that without loss of life. *However, the seabed is littered . . . with millions of tons of explosive ordnance and is the resting place for countless cables, wrecks, moors, nets and junk of every conceivable kind.* Add to these the natural hazards of the sea, traffic, sedimentation and underwater topography and it can be seen that one day, one of the salvage submersibles will itself need salvage assistance . . . The risks to salvage submersibles and their crews are many: the risks will doubtless be faced and accepted with courage. (Italicised words from a paper 'Hazards of the deep' by Busby, Hunt and Rainnie, *Ocean Industry*, vol. 3, nos. 7,8,9).

Three years later, the prophecy came true in every way and the only warming aspect is the truth of that last sentence in particular. Without courage, that *Pisces III* salvage would have been quite in vain. In particular the Vickers organisation spearheaded by Commander Peter Messervy did magnificently in every way and no praise can be high enough for the conduct of Chapman and Mallinson. After the salvage, all submersibles went back to their programmed work without hesitation. And – after a short period of recuperation and leave – so did the two rescued crew members.

It is a facile modern observation that nowadays things are going downhill – implying that they get steadily worse. *All* the participating personnel of this operation showed that long hours, overcoming of adverse weather, dedicated application and wholehearted teamwork are every bit as forthcoming as they have ever been in the past.

So much for the serious side of submarine salvage. On the reverse side of the coin, it was intensely amusing (except to those considered, to whom it was frustrating even if not frightening) when as long ago as 1968 several submarines were salvaged which did not wish to be. On 11 January a trawler netted HMS *Grampus* off Lorient and brought her to the surface. A couple of weeks later, a Corsican fishing boat bagged a French midget submarine in the Gulf of Ajaccio. And then the large French trawler *Lorraine Bretagne* caught the biggest one of all. The 'catch' brought the trawler up all standing as she hooked the nuclear powered Fleet Ballistic Missile Submarine *Robert E. Lee* of 6,700 submerged tons off the Irish West Coast.

It must in fairness be added that when a submarine is in an operational state of nearly neutral buoyancy, it is easier to bring her up when stuck on the bottom! And she wants to surface to see what goes on.

If one writes lightly in fun, there is always an outside chance of some subsequent happening to turn the fun sour. It must therefore be recorded that, since those harmless trawler/submarine contacts, others have ended in damage to the trawls or the trawlers or even in death to crew members. Every seafarer (surface or submarine) will be equally dismayed by such tragedies. The sea is a hard taskmaster –and so is the pen. I cannot **un**write the two unfortunate paragraphs which have been read all round the world in the first edition.

As Omar Khayam put it so succinctly:

'The moving finger writes; and having writ,
Moves on: nor all the piety or wit
Shall lure it back to cancel half a Line,
Nor all thy fears wash out a word of it.

One last anecdote from the past before moving on – the submarine which was salvaged twice. At the end of World War II the *U 2365* was scuttled in 60 metres of water in the Kattegat. In 1956 a well-planned operation raised her primarily for the scrap metal value. After eleven years on the bottom, she was found, however, to be in excellent condition and it was decided to re-fit her for further service. She eventually commissioned into the then-new Federal German Navy as the *Hai*. On 14 September 1966, a welded seam gave way and once more she sank off the Dogger Bank. She was raised again and this time she was scrapped.

Finally, what of the future? Cargo submarines are already something more than a gleam in the eyes of futurists. When one considers, for instance, that the direct route from Alaskan oilfields to Europe passes close to the North Pole and through the North West Passage, there are evident advantages in direct-link submarine operation – except, in the case of mishap, to the salvors who will have to work in hideous conditions, probably under ice as well as under water.

Salvage of nuclear submarines is likely to bring a new and colossally expensive aspect to the salvage world quite soon. It does not need a specialist naval authority to point that out; it must be evident to most thinking people that the anti-pollution lobbies will never want nuclear ironmongery laying around on the sea bottom anywhere, but least of all, close at hand. There has already been the father and mother of an international row about the dumping of armaments and chemicals on the high seas; the sea used to be regarded as the biggest dustbin in the world but is now regarded as the major life-source for the future.

It is equally evident that no navy wants any other navy poking about, investigating, sampling or otherwise gathering intelligence about one of its sunken submarines. So either a permanent guardship on the spot or recovery of the complete vessel will be an essential. The guardship alternative is a high continuing expense, and a drag on naval resources for a number of years with no absolute certainty even then of preventing poaching. The second alternative, of complete recovery, requires an astronomical once-and-for-all expenditure but results in full confidence that no secrets have been divulged.

There will be many discussions in high places about those alternatives plus the ever-popular alternative of leaving the whole thing alone and hoping for the best! As the fairground barkers used to comment so sagely: 'You pays your money and you takes your choice'.

Chapter 9

Aircraft salvage

Having been connected with the salvage of aircraft for many years, I scrutinise each new type with intense interest. Not with admiration at each technological advance, or at the advantages accruing to passengers or cargo operators or to carriers of advanced weapons. Those details are for others. I scrutinise the dimensions, weights, construction and layout of each new arrival with a view to eventual salvage operations. Not the normal outlook but a realistic one I should be banned from international airshows for ever if the organisers discovered I was only visualising the latest creations scattered on the seabed in pieces requiring recovery.

The need for salvage operations arises from many causes. Faulty operation or some omission on the part of the aircraft crew; material failure; weather factors; collision or violent action to avoid collisions; intentional ditchings; engine difficulties; control difficulties; fuel exhaustion; bird ingestion. There are other causes but enough is enough.

As in other types of underwater recovery, salvage cannot begin until the wreckage has been located. If the aircraft was on a scheduled flight, from one airport to another, an approximate position is immediately available. The aircraft may be observed to disappear from a radar screen or it may be in communication with the ground and stop in mid-sentence – those are good checks on timings and positions. Splash witnesses, if available, are valuable even if sometimes contrary in their directions. The position of floating wreckage is useful. After all, or some, of that, detective work it is on with the underwater search as already described for submarines – but the search is so much more difficult since an aircraft is a greatly smaller target and has much less metal to give a magnetic response. Sometimes a device for water sampling to detect the presence and level of hydrocarbons has been used but other methods are simpler and just as productive. SCUBA divers are useful down to about 30 metres depth.

Patience, assiduity and use of every available facility is the only key to aircraft location. Some years ago, when I was in charge of the then deepest salvage operation, it was 31 days before our underwater television sighted the first bit of *Comet* G-AYLP. A little more recently an Eastern Airlines Douglas DC-8 crashed in Lake Pontchartrain; this is a smallish lake 60 by 40 kilometres, 5 metres deep with about another 5 metres of mud silt and took an unremitting search of two weeks before wreckage was discovered. It is impossible to avoid a gloomy outlook as the days pass emptily by and one's shoreside bosses are getting impatient. But it is possible and it is necessary for the top salvage man never to show that he is gloomy.

One primitive method of location is frequently the most effective. On a reasonably flat sea bottom, trawling will locate wreckage more positively than all ultra-modern devices. (The most modern surveying sonars can certainly locate aircraft wreckage efficiently and often provide an excellent picture as an aid to recognition. But pieces picked up in a trawl are very much more certainly and usefully examined by accident investigators). Mark out a probability area around the best calculated datum. Hire trawlers of suitable size and then trawl uptide and downtide through the area so that anything missed on one run may be snagged on the return. With the use of modern precision navigation methods, the trawl should slice through the area systematically like a grocer's slicer going through the bacon.

Navigational equipment may be precise enough but the behaviour of the trawl cannot be known accurately enough to guarantee 100 per cent coverage. But aircraft crashing from great heights usually break into many hundreds of pieces and just one piece in the net shows the search to be in the right place. Then one can reset the datum point, contract the search area and go over and over it again until it comes good.

In the *Comet* salvage we also trawled up the clearly marked tail fin of a York bomber which had disappeared years previously. A chance and 'bonus' discovery which cleared up a mystery which had worried the authorities exceedingly. They had wondered anxiously whether the aircraft had deliberately overflown its destination and been delivered illegally to one of the newly emerging foreign powers. In fact, the unfortunate aircraft had merely perished sadly in an unnoticed accident. Discovery was a most useful side effect of the main operation.

Because an aircraft is easily located in shallow water, it does not follow that salvage is always correspondingly easy. This is especially so in wartime but could equally apply in peacetime military exercise conditions. For instance, on 11 December 1944, a Sunderland flying-boat got foul of the Plymouth anti-submarine, anti-torpedo boom defence and was badly damaged. She did not sink entirely but remained awash. While the Air Ministry salvage party was attaching flotation pontoons, there was a violent explosion which tore the aircraft to pieces beyond recovery and killed five of the seven salvage workers. It was probable that one of the aircraft's depth charges had been dislodged and had fallen to exploding depth.

Up North in the Shetlands, a similar tragic accident was averted by diver Dan Robertson; one-eyed and sixty-years-old, he took brief instructions from an explosive expert, then went down to set the depth charges to 'safe' before starting on salvage work. He was subsequently awarded the British

1 Norwegian Passenger Liner, *Venus*. Drove ashore in severe gale in Plymouth Sound. Refloated by Admiralty Salvage Organisation, after adjusting draught and trim by the use of compressed air, towed to Plymouth within four days.

2 One that got away. The Training Ship HMS *Conway* broke away from her tugs and blew ashore. As can be seen, her back was broken before salvage could be effected.

3 *above* A Mersey Docks and Harbour Board operation. Coaster *Lurcher* being lifted by 'camels'. Note the alarming attitudes of the lifting craft. In such cases if a lifting wire parts, critical new attitudes may be assumed in a split second. It is then 'every man for himself and Devil take the hindmost'.

4 *right* Salvage of MV *Lairdsfield* which capsized and sank on 6 February 1970 spewing her cargo of steel on the seabed. She was raised by the 800-ton floating crane *Magnus III*, the cargo subsequently being recovered by the salvage vessel *Topmost 20*.

5 Tugs from the then United Towing fleet *Welshman* (foreground) and *Lloydsman* attend the crippled tanker *Carlantic* en route for Rotterdam. After going to the tanker's assistance near Sao Thome, off the west coast of Africa in 1972, *Lloydsman* undertook a marathon eighty-one-day tow and finally delivered *Carlantic* to her owners in Holland.

6 The same fleet's 7,500 hp tug *Euroman* pictured off Cape Town with the crippled 29,000-ton Greek tanker *Gallant Coloctronis* in tow. The tanker ran aground on the Mozambique coast in November 1972 and was towed to harbour by *Euroman* in a highly combustible condition.

7 Last stage of *Truculent* operation. Supported by a salvage vessel and pontoons, she is towed to port.

8 A net full of typically fragmented aircraft wreckage, being brought aboard a salvage vessel.

9 Small aircraft brought in by the salvage vessel and now being lifted ashore.

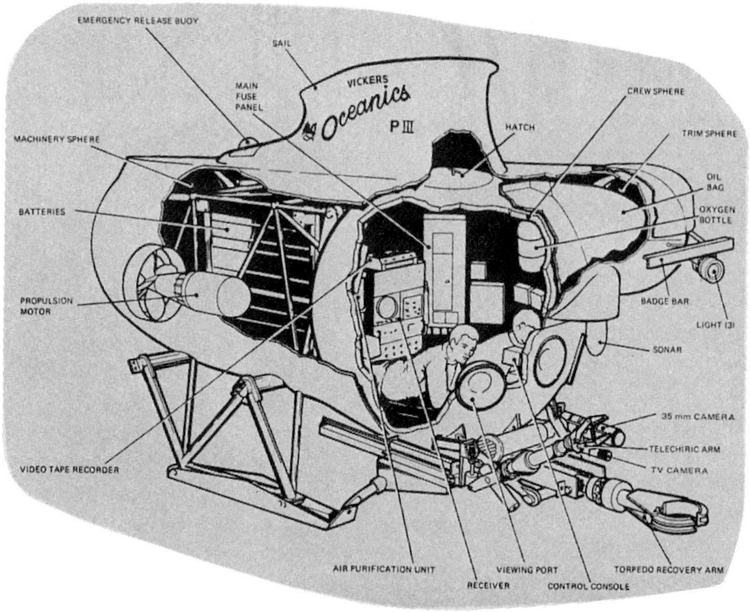

10 Cutaway diagram of *Pisces III* in which two men had to spend 75 hours and 55 minutes on the Atlantic seabed after a mishap. Now long out-moded by more sophisticated craft but a wonderful early submersible.

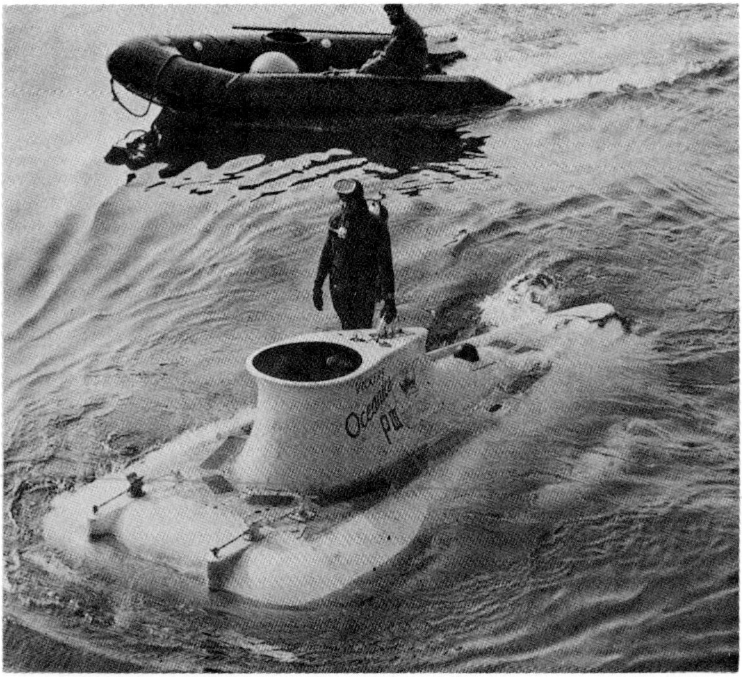

11 Submersible *Pisces III* on the surface between dives. The small size can readily be appreciated by comparison with the *Gemini* dinghy in the background. A pioneer craft which pointed the way to successful development and progress.

L.C. 10.
HARTLEPOOL

12 *above left* French destroyer *Maillé Brézé* sinking in the Clyde after a 1940 explosion.

13 *below left* French destroyer *Maillé Brézé*. All top hamper has been cut away by divers to make the wreck an acceptable weight for lifting.

14 *above* Brunel's *Great Britain* as a bottomed hulk in Sparrow Cove, Falkland Isles. In her delapidated and strained condition, an ordinary straightforward long distance tow was out of the question. The *Ulrich Harms* pontoon was therefore carefully sunk, the ship positioned over it and the whole unit refloated.

15 *above* *Great Britain*, on top of the pontoon being towed home to Bristol for restoration and exhibition.

16 *above right* It takes all sorts to make a salvor's life. Here are some of many live mines recovered from a long-sunken minelaying vessel.

17 *below right* During World War II these were part of a complex of anti-aircraft forts in the Thames Estuary. Long after the war it became a harbour clearance task to cut them down and remove them leaving a clean seabed. Note the strong tidal conditions (bottom right of picture) which made the divers' work very difficult.

18 A cactus grab bringing wreckage from the seabed 200 metres below.

19 An Italian type of observation chamber from which an observer directed operation of the recovery grab.

20 An 'Iron Man' of pre-World War II days; note the claws for working with. Not an operational success.

Empire Medal – and very richly deserved it because the charges were within a very small and touchy distance of their set exploding depth. He was a very brave man.

Probably the Air Ministry's greatest success was when a KLM Super Constellation crashed on a mudbank in the Shannon Estuary on 5 September 1954. With the aid of rubber lifting 'camels' and with much improvisation and ingenuity the whole broken up mass (with the exception of one deeply buried engine) had been recovered by 20 September. The recovered cargo included £400,000 worth of diamonds, cases of waterproofed watches and registered mail.

In passing it is well worth notation that with the post-war shrinkage of the Services, and in particular the passing of the flying-boat era, the Air Ministry Moorings and Salvage Branch was closed on April 1 1966. The Directorate of Marine Services (Naval), of which I was then Deputy Director, took over mooring and salvage responsibilities for all three armed Services from that date. It was a sad disappearance of a versatile and successful Air Department unit but on the whole it made good sense to rationalise and above all – dearest to the hearts of every politician and Service chief – it was more economical. It was an inevitable step.

Now let's look at the final above-water incidents before passing to the depths. Both are excellent examples of the ingenuity and improvisation essential to salvage success.

The first of these incidents again concerned one of those ubiquitous Sunderlands. In 1954, it ran aground at Britannia Lake, Greenland, when transferring equipment there from the British North Greenland Base at Zackenberg. The Sunderland was initially refloated by means of placing a collision mat over the damaged part of the hull and by removing water from inside. However the flying boat was not riding nearly high enough to be able to carry out plating repairs to make the aircraft serviceable and facilities were dreadfully sparse in Greenland. The ingenious aspect of this case now came with the use of a twenty-man inflatable liferaft. It was spread on the water while deflated and then manipulated under the damaged section. The top of the liferaft was stiffened and strengthened by use of radio-mast tube sections and the raft was inflated. This novel improvised lifting craft worked splendidly, raised the aircraft high in the water and enabled satisfactory repairs to be made in the dry.

The second incident around a year later concerned a DC-4 of US Overseas Airlines which had to make a forced landing on an icefloe in Hudson's Bay. Sawdust and ice were used to prevent the floe melting. A helicopter put a salvage party on the floe and supplied them with equipment and food as requisite. Then the whole *ensemble* was towed by local craft to Churchill. The aircraft used its own power in partial assistance to the necessary small and improvised towing force available in that remote area. Ice is normally a terrible enemy in Hudson's Bay but on this occasion clear thinking and commonsense turned it into an ally.

In most cases the salvage of submerged helicopters is easier than that of fixed wing aircraft. The latter as already remarked enter the water at high speed and fragment themselves; the former are inclined to flutter in at low speed and in gradual descent and are thereby usually found in one piece. There is generally a good strong lifting point available on a helicopter and it can often be lifted as a recognisable entity. I quote one typical operation below. By omitting names and dates it becomes a blueprint for similar operations so far as can be seen in the future. As a bonus it shows that despite faultless planning – and as the fault of nobody –something untoward will almost inevitably retard *any* operation.

A naval helicopter ditched some six miles south of Portland Bill and to establish the fault, salvage was ordered. A naval mine counter-measure vessel (MCMV) located, with her minehunting sonar, a conspicuous and promising target on the seabed in approximately the correct place. That done, she vectored another vessel into a position immediately over the contact so that a buoy marker could be dropped exactly alongside the target on contact. The next step was for a

Scuba diver to descend and identify the wreck; luck was in and this was done quite quickly despite rough water, 55 metres depth and, because of sand in suspension, nil visibility down below.

Two divers next took the lifting wires down, but due to a misunderstanding, one diver believed the other to be in trouble and rightly brought him to the surface; as a consequence the end of the wire was left hanging in the vicinity of the helicopter but not attached to it. A fourth diver now went down to connect up but the tidal stream was beginning to run so strongly that it proved impossible to do the job properly; however, correctly assessing that *any* firm connection to the surface to be better than none, this diver managed to secure the wire to a wheel bracket before he was forced by the tide to surface.

The good luck so far was beginning to run out and at this juncture the weather went sour and it was impossible to work again for a period of three days. Then, again in nil-visibility and working in dangerously jagged wreckage, divers managed to heave the wire over the top of the helicopter, which was on its side, and down on to the rotor head which was against the sea-bottom. The helicopter was then winched to the surface by a 'horned-type' salvage vessel and carried to the harbour.

It was a good job done in quick time despite three days bad weather. Had the divers not misunderstood each other on the first day, something else would have gone wrong. It is axiomatic that no salvage operation can proceed without some spanner in some works at some time.

Just occasionally when a pilot has carried out a carefully controlled sealanding, a fixed wing aircraft may also be found in one piece but such occasions are pretty rare. Much more often the aircraft has plummeted with all hands after mid-air disaster or – mostly in the case of military planes –the aircrew have been able to eject and leave the plane to come down on its own. The G-ALYP *Comet* off Elba was an instance of the disaster type. It is useful to describe that operation in synopsis form because it was an historic first-off job which taught lessons and established the principles firmly in use today.

On 10 January 1954, disaster struck the British Aircraft Industry. A *Comet* airliner, on which so much of Britain's air hopes had been pinned, plunged into the Mediterranean Sea off Elba. Because of the importance, the British Naval Commander-in-Chief Mediterranean (Admiral Earl Mountbatten of Burma) immediately contacted the Minister of Transport and Civil Aviation (Mr. Alan Lennox Boyd). They jointly decided that every effort should be made by the Royal Navy to salvage the aircraft so as to carry out a thorough investigation. The Minister promised every possible assistance, including financial, to the Navy.

Location of the aircraft proceeded in the now classic manner. An instantaneous cut-off of a radio conversation with an airport was a pretty sure indication of the time and rough position on track. Next the area where bodies and floating wreckage had been discovered and picked up was another possible pointer. Unfortunately the picking up boats were simple fishing boats, were not equipped with navigational aids and had not noted bearings or times; so the crews had only the roughest ideas of the picking up positions. Eye witnesses were found and interrogated; eleven reliable witnesses were selected and the remainder rejected. The captain of another passing airliner heard of the problems and sent along a photograph he had taken of the fishing boats at their rescue work; that gave one very good line of position. Gradually a probability area emerged so that the next stage could take place.

HM ships combed the area with sonar and dropped buoy markers alongside likely contacts. The authorities already appreciated that even a complete aircraft would give a poor sonar response but no one could foresee what would actually be detected. We found an old old wreck, a heap of antique amphorae, a lot of Italian wartime mines, some sunken buoys and eventually some pieces of *Comet*. Each had to be laboriously marked by dan-buoy and each had to be laboriously identified before passing on. The depth of nearly

200 metres meant, of course, that no ordinary diver could investigate and that everything had to be looked at by observation chamber or television. It was a fiddling, frustrating, time-destroying job.

Television equipment and observation chamber were carried in the salvage vessel *Sea Salvor* together with a ten ton grab to recover wreckage. To investigate each contact, it was necessary to position the salvage vessel precisely and vertically above the 'target' – and because of the great depth of water, and the prevailing winter weather, the vessel had to be moored there very securely indeed. The type of mooring then favoured had six 'legs' spread on the seabed at 60 degree angles apart; each leg had anchor, cable, wire and buoy which made it about ten tons in weight. Shifting six of these was a long business even in fine weather but the weather was more often horribly adverse and the work heartbreaking. But because of the number of contacts to be investigated, these moorings had to be shifted more than a score of times. The salvage crew was magnificent.

To augment all the scientific and specialised equipment, a small fleet of trawlers was hired. A naval officer instructed them in elementary minesweeping techniques and in station-keeping on each other for best overlapping results; although ragged at first, they became quite skilful. A trawl has two advantages; it will locate items for subsequent identification and, if they are not too big, it can also bring them to the surface. In the closing stages of this operation, when all the heavy wreckage has been lifted, trawling was used here to the exclusion of all other methods. It produced good dividends, possibly 10 to 15 per cent of the total recovery.

Weather, as always, was the worst enemy of the salvage. For the first two months, gales followed gales. When it was not blowing, the residual swell made work almost impossible. There were other enemies too. Merchant ships, passing at night, rammed marker buoys and sank them. Fishing boats dragged the markers out of place with their trawls. Worst of all, a large seagoing tug brought her tow through the area. With the long catenary in her towing wire acting as a perfect sweep-wire, she cut seven markers adrift in a few minutes. The area had been broadcast to all shipping as restricted but on the high seas one can only *request* vessels to keep clear. There is no legal authority for issuing prohibitive orders.

Despite all these difficulties hampering work, success came in the end. First in small pieces of fuselage, personal effects, the after pressure dome and passenger seats from the rear of the aircraft. Then the main wing sections followed by four engines and most of the undercarriage. Last, after another prolonged search, the forward end of the aircraft. This was the biggest lift of all and included the whole flight deck with controls.

All those big recoveries were by controlled grabbing. A similar type of grab can be seen working on coal heaps or discharging bulk cargo from ships. It is lowered or raised on one wire and is adjusted by another. The second wire opens or closes the claws. It is like an enormous eight-fingered hand. In spite of its apparent clumsiness, it does little damage when guided expertly by a 'diver' in an observation chamber near the seabed some 200 metres down. The 'diver' looks for a good strong component on the wreckage and directs the grab handlers aboard the salvage vessel on how to close the grab over that part. It is no exaggeration to say that it could pick up an article like, say, an ordinary sized coal scuttle without damage. On the other hand, it will take up a 10-tonne weight equally capably.

The success of such salvage enterprises is invaluable. The *Comet* wreckage was, for instance, given intensive testing at the Royal Aircraft Establishment, Farnborough. Experts 'rebuilt' the plane for metallurgists, physicists, designers and other backroom boys to work on. The cause of the accident was eventually attributed to metal fatigue after thousands of expansions and contractions in various air temperatures and pressures. Metal fatigue had, of course, been known to occur but had previously never been considered significant. As a result of the salvage, the *Comet* went on to great triumphs over the years. Perhaps not the least of those triumphs is that 35 years later modified *Comets* renamed as *Nimrods* were still an important part of Royal Air Force operations.

Should such an important salvage operation be required nowadays, in similarly circumstances, it would be carried out much more quickly. The location procedure is now well practised instead of being formulated on the spot. Modern navigation aids would make the loss of markers inconvenient rather than disastrous. A vessel capable of maintaining a precise position on the sea surface, (by means of angled propulsion units, sensors, and computer control i.e., the so-called 'dynamic positioning') would be used instead of all those cumbersome wires, chains and anchors. Acoustic beacons laid on the seabed would be an efficient modern substitute for surface marker buoys. Possibly a submersible would take the place of the observation chamber. Ships themselves are faster in getting to and fro. And helicopters, by ship-shore transfers of mail, personnel and stores can keep the ships longer on task without necessity to break off in the middle of operations.

All those helpful items are for the using if money and equipment is available without penny-pinching. On the other hand, if it is a question of make-do-and-mend with readily available resources it is different. Then one might as well brush the dust off the *Comet* salvage report and do a carbon-copy operation.

Still deeper salvage is operationally possible but extremely expensive. In that case, the use of a submersible to take photographs or to recover the 'black box' may well suffice to establish the cause of accident. There was an early pointer towards this in 1969 when a Scandinavian Air Lines System DC-8 crashed in Santa Monica Bay approaching Los Angeles airport on 13 January. In fact the depth of that wreckage was only 115 metres. But the submersible used, *Deep Quest* is capable of operating to a depth of 2,400 metres – so the extension of such an operation downwards became evident from that time onwards.

The salvors of that DC-8 had particular good fortune. The aircraft broke into three pieces on hitting the water. Two pieces sank but the other piece remained afloat long enough to be towed into shallow water where it was easily accessible and recoverable. Thus the wings, forward cabin and cockpit were recovered rapidly. The large, floating portion of wreckage had also enabled the salvors to know the position of the sunken portions with fair accuracy and the effort expended in location was practically nil.

Headquartered in nearby (by American standards anyway) San Diego, the Lockheed Missiles and Space Company had been carrying out deep trials with the submersible *Deep Quest*, a sophisticated little vessel launched eighteen months previously. This submersible was therefore assigned to the task of looking for and recovering the DC-8's flight recorder. Exactly one week after the accident it descended and found wreckage in the very short space of one hour. That customary bugbear, the weather, then intervened and drove everyone off task for two days.

On 22 January *Deep Quest* dived again in search of the tail section containing the flight recorder; this was quite speedily accomplished by sonar and shortly afterwards by eye. An attempt to shift the tail section was aborted by the construction of the elevator which did not permit the particular manipulator tool available to get a grip. However, on 4 February *Deep Quest* found and recovered the flight recorder as a grand finale to a most useful three-week mission of photography and visual examination by accident experts.

As a point of further interest, *Deep Quest* carried out an almost identical operation in January 1971; curiously enough, it was in the same bay but this time on the edge of the offshore slope at 330 metres depth. Although this project was made more difficult by worse weather and deeper water, there is no advantage in giving a blow-by-blow account as it would duplicate the previous account in too many respects. Suffice to say that *Deep Quest* found and recovered the flight recorder again – and proved her capabilities 215 metres further down than the recovery two years previously.

So much for official investigation and recovery. What would induce a commercial salvor to enter the running? Well, I give just one example. Very recently there was a Press report that an airliner bound for New York had suffered engine trouble over the Atlantic but had just managed to struggle back to London and land safely. In the cargo was a consignment of diamonds worth £3 million. Had she failed to make the airport and landed in the sea instead, things would have really started to happen. Salvage firms world-wide would have been doing lightning financial calculations and deciding whether to employ their highest-power lobbying to get the job. Even if there were prospects of a slight financial loss, the resultant prestige and increase of professional standing might well have decided them to volunteer for the new style task.

Every now and again, there is another and more sombre reason for recovery to take place. Whenever an airliner goes down with a VIP or, say, a group of officials on their way to a conference, the first thoughts of official and journalistic minds are of possible sabotage. Usually this first thought proves to be incorrect. But, on two occasions in my experience, sabotage has eventually been found to be the most probable cause. In one of those cases, a difficult deep recovery was necessary. In the other case, the waters were shallow but shark infested. There is virtually no easy case in aircraft salvage.

In a later chapter, I shall discuss the virtues of different techniques of work and recovery at various depths. But it is worth saying specifically in this aircraft chapter that, in this sphere of activity, it is usually the most practical seamanlike and least scientific methods which seem to bring in the best results. If the aircraft is fragmented (as it so very often is) and the seabed is at all suitable, plain honest trawling will always be best of all. Trawls will collect small and medium-sized pieces without difficulty. And large pieces can be snagged and marked for other types of elementary 'pulley-hauley' methods of recovery. I have seen three-quarters of a tonne of aircraft tail come up in a trawl. I have likewise seen a photograph of a one-tonne boulder which, on being surfaced, gave no one any pleasure or sense of achievement!

The precept for recovery trawling is akin to that attributed to Robert the Bruce. 'If at first you don't succeed, try, try and try again'. Trawl with the tidal stream, trawl against the tidal stream and, if there is sufficient period of slack water, trawl cross-ways. It is surprising what treasures different approaches to the very same spot will produce. And if the accident investigating officers complain – as they invariable do – that one or two vital clues are missing, it is well worth going back for a repeat performance a week, a fortnight or a month later. The sand in the North Sea, as one example, is like a magic carpet. A savage and prolonged blow from one direction can well cover wreckage with a couple of metres of sand and make location or recovery impossible. But a savage and prolonged blow from some other direction may equally well leave wreckage standing once again quite clear of the bottom. Sometimes the wreckage, especially the lighter material, may have travelled miles from the original position.

It is safe to say that accident investigating officers always start by pleading for *anything* and always finish by demanding everything. I say again, *everything*. Fortunately there is always some hard-headed person at the top who will decide when the financial outgoings on the operation are not being sufficiently repaid by incomings of salvaged material. That calculation is the only factor which enables any salvage officer to terminate any aircraft recovery operation.

As in the case of submarine salvage, lifesaving must take precedence over salvage. Only the time scale is shorter. A great many people have survived ditchings. But I remember, with particular pleasure, one wonderful rescue in the Pacific. A Pan-American Stratocruiser made an intentional forced landing because of propeller failure and it took twenty-one minutes to sink; during that short time the US Coast Guard cutter *Pontchartrain* approached and picked up seven crew members and all twenty-five passengers.

I close with two stories about the dearly loved flying-boat service – dearly loved in all parts of the world as slow, safe, reliable, land-anywhere aircraft. I made my own first flight not in a pure flying-boat but an 80-knot amphibian – officially known as 'Walrus' but unofficially known as 'the steam pigeon'. One of the greatest advantages of the boats and the amphibians was that when in trouble they could land on the sea (a contradiction in terms, but that's what they called it!) and float until assistance arrived.

First there was an autumn day back in the fifties when the SS *Harry Culbeath* found an immobilised Catalina on the sea 530 miles south-east of San Francisco. Without blinking an eyelid, she hoisted it and the four crew members aboard and carried it into the harbour. Salvage without tears – the only example I have ever known.

Second and finally, in September 1954, a Sunderland flying-boat was carrying an RAF football team to Jersey; unfortunately when alighting it struck an underwater rock half a mile offshore and started to fill up. It taxied halfway to port and was towed the remainder by a miscellaneous collection of available craft. She was then beached just like a ship salvage case, and at low water a 10 metre slit was boxed in, also like a ship salvage case, with a good old-fashioned cement box. It was fortunate that the RAF football team was on board; the players were immediately employed as 'volunteers' to make up the salvage working party. In the Services, it doesn't take long to become a volunteer!

TAILPIECE — This extract from the *Washington Post* of 5th September 1930, may describe the first-ever salvage of an aircraft from the sea:

SUBMARINE LIFTS SUNKEN SEAPLANE

(From Our Own Correspondent) — Paris Sept. 4.

An extraordinary feat of salvage was achieved near Toulon yesterday.

A Naval seaplane was forced down to the sea by engine failure and was about to sink, its crew having been rescued by fishing boats.

The submarine *Romazotti*, which was near on the surface, dived and came up under the plane, which it lifted out of the water and transported on its back to Toulon. The officer commanding the submarine has been warmly congratulated for his ingenuity and daring.

Chapter 10

Medium and deep water recoveries

There has already been mention of deep recoveries in the submarine and aircraft chapters. But there are deep operations in several other spheres and the object of this chapter is to bring them together.

It will be a good idea to open the subject with a present day definition of shallow, medium and deep. Naturally as many opinions exist as there are commentators. In my opinion, however, the finest practical dividing line comes at the point where ordinary/average divers can no longer perform *significantly useful* manual work – currently no deeper than about 150 metres. Down to that point may nowadays be termed shallow. The next practical dividing line comes where recovery by trawling and by directed grabbing is no longer practicable as a means of retrieval. Although trawlers do operate to depths of 550 metres, or more, the vessels are a bit big and the gear too heavy to be responsive enough in 'feeling' for wreckage. A more workable depth is 450 metres. So from 150 metres depth to 450 may be regarded as medium. And anything below that second line is therefore, 'deep'.

Naturally there is no law or objection which forbids using deep or medium techniques in shallow water. Nor against using deep techniques in medium depths. If the equipment is readily available, the salvage officer will wish to use the one which will cope with the sutuation best and fastest. Efficiency and economics are the only arbiters. But the important thing to emphasise is that limits and dividing lines move remorselessly deeper as the years roll by. And they have quite recently moved down remarkably rapidly, for only half a century ago 200 metres was considered mighty deep indeed to carry out any sort of work.

I hope, therefore, by a brief run through various past cases to show how the depth barriers have been pushed steadfastly downwards case by case. It is a fascinating story.

As already mentioned in chapter one, at the outset of the Christian era, recovery of sunken property, as recorded by Livy, was recorded strictly according to depth. Divers recovering items from 8 cubits (4 metres) entitled the salvor to one-third the value; recovery from 16 cubits (8 metres) entitled the salvor to half the value. Since no values were mentioned for greater depths than 8 metres, evidently that was considered as near the extreme deep limit. There is, quite obviously, no time in this small book to do more than mention that early point as a starter – and then to make a great leap through the centuries towards the present. From Livy's time to post-World War 2, the pushing downwards was significant but far from startling; the subsequent years has seen a revolution. There is just no other appropriate word to use.

Let's first go back to 1917-24 when the years-long salvage of bullion from the sunken White Star liner*Laurentic*, with a seabed depth of about 42 metres (call it 84 cubits!) was considered quite phenomenal. The vessel had been sunk, in January 1917, by an enemy minefield off the mouth of Lough Swilly on the north coast of Ireland. Even by today's standards it was a formidable task – not so much by depth but because of the position being extremely exposed to heavy weather. When the Atlantic weather was from west or north, the salvors took the full relentless force; when from the south, which might have been expected to provide a lee, the 20-mile 'tunnel' of Lough Swilly gave the wind and sea a venemous drive. But an attempt at salvage was essential; the vessel had been carrying bullion worth (at 1917 prices) £5 million and the money was much need for sustaining the all-out war efforts.

The wreck was found lying on her port side at an angle of about 60 degrees – a particularly difficult position as it meant there was no flat platform anywhere for a diver to work from. In the tremendous back and forth surging of the sea in that exposed position, a working platform would have been a great asset but, that not being possible, work had to progress with divers hanging on for dear life with one hand and working as best they could. A splendid example of that old evergreen seaman's adage: 'One hand for the King (or Queen) and one hand for myself'. There was, of course, as might be expected in a hastily abandoned ship plenty of loose equipment also surging to and fro under water, in search of a victim to entangle, stun or otherwise damage. Altogether a horrible task.

The position of the gold within the ship was of course known to certain officers of the *Laurentic* and to certain officials of the White Star Line. Thus it was known that if entry to the second-class baggage room could be effected, salvage was at least going to be worth a try. Ship's plans and drawings were available and so salvage personnel could be briefed thoroughly on dry land beforehand on what they might expect to find underwater. Armed with this information, the salvage vessel took up moorings and stationed herself over the wreck. The divers descended, blew off ship's-side entry doors and had them hoisted out of the way aboard the salvage vessel. Once inside the ship, a security gate, certain other cargo in the way and a storeroom door had to be dealt with by a mixture of explosives, hard physical graft and cutting tools.

As ever, the process seems easy in retrospect. But, operations being punctuated by severe winter weather, the arrival of divers at the strongroom took nearly a fortnight. Then, in a very confined space, Diver Miller got four boxes, each worth £8,000, hoisted to the surface. The way ahead looked good – wreck located, entry effected and first recoveries made. But salvors know too well never to count any chickens before they are hatched. A week's hard gale blew from the north, the resultant sea and swell partially collapsed the wreck, the entry port was located 13 metres deeper than previously and passage to the strong rooms was barred by crushing of the ship so that roof and floor of the corridor were only half a metre apart. A way was blasted through but on re-arrival at the strong room, it was found that much of the gold had spilled out on the seabed through the ruptured hull.

Because of the collapsing stresses on the ship's structure, the route to the strongroom had now necessarily to be considered too dangerous for personnel to work in; the most practicable albeit infinitely tedious, method was to dig down vertically, by cutting and by controlled explosions, from topsides of the wreck right through her to the strongroom and seabed. There was no easy way and each piece of obstructive metal had to be hoisted clear by the salvage vessel and stowed aboard for subsequent dumping well away from the working area. As a result, it took some two months to get in touch with any gold again. Then came the good part. By September 1917, a total of 542 bars out of the 3211 bars sunk had been recovered. Then, aggravatingly, the salvors were switched to another more urgent project and did not return for eighteen months.

In spring 1919, the work commenced as before and, at first, the recovery rate remained good. Then despite the many bars known for certain to be there, the discoveries began to run out. Nevertheless, some £470,000 worth was pulled up that season. At the re-start in 1920, the intervening winter weather had again broken the ship and changed her configuration. From then on, unremitting hard graft was required to blast, cut and clear away seemingly interminable obstructions – and added to the ship's own material now was a huge amount of sand and stones washed in from the outside. It was a tedious heartbreaking process with only a very occasional bar of gold turning up to give encouragement. In fact, only seven bars came up that season. Next season was only slightly better with forty-three bars. Then, at last, the luck began to turn again.

As a result of winter 1921-2 gales, much silt had been washed clear, the ship became still further ruptured and the Spring resumption welcomed the sight of gold actually *showing* itself again. That bright year produced 895 bars, the next year (1923) recovered a record 1,255 bars and finally (1924) the total was 129 bars. That left only twenty-five bars on the bottom and it was not financially worthwhile to mount another season's expensive expedition on the offchance of finding these scattered few.

Thus ended an epic recovery operation where the luck ran good, then bad, and finally good again. Lieutenant Commander G.C.C. Damant was responsible for the whole operation from the salvage vessel *Racer*. He was promoted to Commander towards the end and rarely can an honour have been so richly deserved. Nevertheless, there was one wry twist in the end of the story – almost unbelievable but entirely true. Although promoted, he could not be paid as a Commander because the complement of the *Racer* did not allow such a high ranking officer to be borne aboard for duty. He had either to continue on Lieutenant-Commander's pay or give up the *Racer* on which he was the supreme expert. The Sea Lords of the day lent all their weight to bend and re-interpret the rules and eventually all came right. But considering the enormous benefits Damant had presented to the Treasury, it was a sadness that such a parsimonious attitude should have ever intruded.

Modern salvage officers who occasionally fancy that the office administrators love rule-books, better than good results can take heart and know that it was ever thus. They are most certainly not the first and equally certain not the last to fall foul of the abominable 'no-men' who exist in every type of business concern. I mean those negative thinkers, so difficult to shift from their entrenched rule-book positions until blasted out by the top men who have achieved their eminence by the exercise of commonsense and regular production of good operational results.

Besides the successful recovery of so much of gold from the *Laurentic*, the side benefits were great indeed. The depth of 42 metres had originally been deemed a tremendous barrier to the efficient working of divers. But, as the years progressed, many divers began to regard that depth as near-routine. Diving techniques were significantly improved, lessons were learned and new de-compression policies were evolved. In short, although perhaps the participants did not themselves realise it, the way was being opened to go still deeper.

During the time that the *Laurentic* gold was being recovered, the Peninsular and Oriental vessel *Egypt* sank after a collision with the *Seine*, a French cargo steamer, about 25 miles south west of Ushant. As can be seen by a glance at a chart or map, that position is about as weather-exposed as any in the world. The swell can run in without the slightest hindrance from as far afield as Greenland, Newfoundland or Brazil. The date was 20 May 1922. *Egypt* carried an amount of gold and silver in coins and bars worth over £1 million in total. Nevertheless, despite that attractive cargo, mainly because of the unfavourable position and a sunken depth of around 130 metres, the wreck lay there undisturbed for nine years; then the Sorima Salvage Company of Italy undertook to recover the gold and silver.

In May 1931, after formidable difficulties of location and identification, the Italian salvage vessel *Artiglio 2* moored over the top of the wreck on a six-point mooring. The job was far too deep for divers and a manned observation chamber was used for directing the placing of explosive charges and for directing the cactus grab accurately enough to pick things up. As in the *Laurentic*, the salvors 'blew' their way down from the topdeck to the strongroom with explosives but, in this situation there was need for even more exactness and care – because of the damaged deck of the strongroom itself, a false or rash action might cause the gold and silver to drop through into the hold underneath and thus present an almost impossible task of recovery.

In passing I must mention one point of particular technical

interest. Before adopting the observation chamber as the main weapon of recovery, the salvors had given a fair trial to an armoured diving suit known as the 'iron man'. It was constructed by the experienced firm of Neufeldt and Kuhnke of Kiel and was the then the most modern of a long historic line of attempted armoured diving suits. It was a grotesque metal contraption looking like a blown-up human being. The metal shell was so jointed and watertighted that the man inside should have been able to move his legs for walking and his arms and hands for working. Unfortunately, it did not turn out to be as flexible as intended.

That fact was that the air pressure inside the 'iron man' remained always at one atmosphere but the water pressure outside increased dramatically as the chamber was lowered deeper and deeper; the joints therefore became extremely stiff and they made it almost impossible to perform any useful work. The *Egypt*'s salvors quite soon abandoned use of the 'iron man' in favour of the conventional observation chamber. Hope, however, springs eternal and I see that another armoured suit is undergoing sea trials as I write this book.

After the previous examples it is not necessary to recapitulate the problems of working in weather-exposed positions, the wind, sea and swell; all these things plus the dangers of using explosives and the additional dangers of working with a manned observation chamber *inside* a damaged wreck while suspended from a surging, heaving, ship on the surface. Something of the problem has been explained in the *Laurentic* and other tasks and anyone who has been to sea in bad weather will be able to use his own imagination about the difficulties and hazards of working from such an unstable suspension point.

Suffice to give one example. In the whole of the 1932 season, 'divers' (in the observation chamber) could manage to get on the actual job for no more than 188 hours total. Nevertheless by dedicated application and enthusiasm, *Sorima* recovered in three seasons' work (1932-3-4) most of the sunken treasure. There was one additional problem. Loose sovereigns could obviously not be picked up by grab so an ingenious 'vacuum cleaner' was devised. It took the form of a strongly constructed tank with a bottom opening sealed with glass. When the glass was intentionally shattered by remote firing of an explosive charge, water rushed in to fill the vacuum, carrying with it any gold coins in the vicinity.

That was a Heath Robinson affair that few people would have considered feasible but it worked well in practice! The whole pick-up concept was a well conceived, planned and executed operation which at least for several years put the Italians right at the top of the league for deep underwater recovery work.

One more example only, of a few years later, will illustrate as well as anything the constant but still slow battle to get deeper and deeper. Then we can jump ahead a decade and a half to see how the slow battle against the constant enemy of increased pressure with depth became, or started to become almost a rout.

The *Niagara* was a World War II overseas casualty; on 19 June 1940, she was mined and sunk off Whangarei on the North Island of New Zealand in 140 metres depth of water. Amongst other cargo from South Africa to Vancouver, she was carrying 10 tons of gold bars worth approaching a value of £2·5 million. Again this was an important recovery required for helping to finance the war and salvage was accordingly carried out under dangerous war conditions in an enemy-mined area. The location phase was long and anxious; in those pre-electronic (well almost!) days the probable area had to be laboriously dragged across with wire sweeps for nine weeks before positive contact was made. The vessel was found with an angle of list 70 degrees from vertical.

Access was, as in the last two cases, effected by blasting and pulling the blasted debris clear. The great angle of list made the work harder and much care was required so as not to break up the bullion room itself and thereby lose the bullion into some much larger compartment underneath. At that depth, of course, all ordinary diving work was impossible and all direction was from an observation chamber, all pulling and hauling was by heavy grab and all firing of explosives was by remote control. It was a fiddling and irritatingly tedious, non-productive task until at last the bullion door was reached. Then that had to be most carefully blown off its hinges so as not to disturb or distort the inside structure of the bullion room. But in due course, on 13 October the first bar of gold was brought up. Just two days under ten months from when the salvage operation had commenced on the previous 15 December.

In under one year, all gold was recovered except for a negligible amount worth some £135,000. The record working day was on 11 November when £350,000 was picked up between 08.20 and 17.20. Which just goes to show that salvage firms do not have it bad all the time; they just remember the bad patches more vividly than the good. Everlasting credit must go to the Australian salvors – Captain J.P. Williams, Captain J. Herd and Chief Diver, Mr. J.C. Johnstone. They were superb. Plain superb.

The *Comet* aircraft salvage, as already related, was considered in 1954 to be an advance in depth capability with recovery from depths down to 200 metres. But aircraft salvage being a technique of its own, it is probably fair to regard the salvage of a tug by use of a submersible from 200 metres as a step downwards in its own right. In 1969, the Vickers *Pisces I* descended in Howe Sound (near Vancouver) and worked on a sunken 95-ton tug *Emerald Straits*. The job was noteworthy as a first of type.

After first cutting away cable securing gear from the *Emerald Straits*, the *Pisces I* went on to remove the casualty's anchor and cable from her hawsepipes so as to make a secure holding place for insertion of a strong lifting toggle. The bows of the tug were then raised by the attendant vessel and salvage lifting wires were guided into position by the submersible. The tug was then lifted and carried to shallow water to enable easier underwater working to be carried out. She ultimately came to the surface after only 100 hours of diving time by *Pisces 1*. To refloat a ship (albeit a very small one) from that depth was a real milestone in the progress of salvage techniques.

From 1969, we move back three years to a much deeper but again very different and very specialised 1966 operation. It is an excellent case to discuss as the search for, and recovery of, an atomic bomb in the Mediterranean off Palomares, Spain, was probably the best documented deep recovery operation in history. Let's get the scene for a start. A B-52 bomber was refuelling from a KC-135 tanker aircraft in mid-air during the forenoon of 17 January. There was a collision and most disastrous explosion and then wreckage from both aircraft showered down out of the sky.

Within twenty-four hours of commencing search, three nuclear weapons were located on land by efficient search teams but the fourth one known to be carried could not be found after a most intensive hunt. It was then assumed correctly that number four had fallen into the sea and the United States Navy was therefore summoned to mastermind the operation under the terms of United States joint services instructions. Four crew members of the B-52 had parachuted to safety but all personnel from the tanker aircraft was lost. So, although not overly valuable to the salvors, there were survivors to be interrogated.

It is evident that such an operation can never be a civilian task so it is extremely useful to look at the naval organisation which was set up. No doubt it will be refined in the light of operational 'wash ups' and by a continuing study. In essence, however, it was a good sound organisation which worked well and produced a thoroughly successful outcome. There have some subsequent criticisms of oversupply of men, ships, and material. But it is my long personal experience that it is *always* better to oversupply than undersupply.

As the wise old United States General, Nathan Bedford, liked to observe, the first recipe for victory is: 'to get thar fustest with the mostest'. That is one more precept which ought to be engraved on every salvage organiser's heart. Surplus supply can be returned with little or no trouble; one missing item can, on the other hand, foul up an operation entirely if not there when required at a critical stage.

In Washington, a Technical Advisory Group was set up

under the control of Rear Admiral L.V. Swanson. Without being presumptious to my elders and betters, I must say what a good move that was. It is always extremely important to have an efficient rear-link headquarters to interpret the intentions of seamen on the job to well meaning but lay politicians (and to interpret vice versa), to keep VIPs and Press off the backs of the operational commanders, occasionally to offer advice from a cool detached position and to organise material and personnel for the task.

Afloat, in the Mediterranean, Rear Admiral W.S. Guest, assumed command of all ships and vessels in a specially created designation of Commander Task Force 65; he flew his flag in USS *Macdonagh*. That is good thinking again. Having a Rear Admiral at each end is one particular good way of getting unorthodox and urgent actions implemented promptly with the minimum of obstruction and dissent.

Every possible support was given from Washington. Four submersibles of varying capabilities were ordered to the scene, as was high-definition bottom scanning sonar, a Hi-Fix navigation unit, underwater television cameras, two survey vessels and the finest available mine-hunting sonar. The United States armament and fuel depot at Cartagena was used as a reception and relaying post for all material. The Spanish Air Force kindly provided the nearest possible airfield for transits. With all the care and thought and planning, logistic needs presented the minimum of problems.

The location phase was far more comprehensive than anything described before. Diving teams searched the shallow waters adjacent to the shore line by the 'jackstay' method; by feeling and looking along a pattern of stretched out wires so that no gaps could be left. Those inshore teams picked up some 50 tonnes of aircraft debris but no bomb. In water depths greater than 40 metres, the search was carried out by bottom scanning sonars in patterns monitored by high precision navigation systems. Contacts pointed out by sonar were investigated by divers or submersibles according to the depth. One of the survey vessels carried out a detailed survey of the area. Another worked an underwater camera over an area pattern.

The contacts were examined and discarded one by one. But on 13 March, the submersible *Alvin* at long last sighted the bomb on a steep sloped bottom with a thick covering of loose shifting silt at a depth of (I quote the actual depth, reported at the time in feet, for accuracy) of 2,550 feet (780 metres). Another submersible, *Aluminat*, was despatched to meet *Alvin* down below and to place a sonar transponder alongside the bomb so that it could never be lost again. In fact that last action turned out to be a vain hope because the bomb did quite quickly get lost again. At the first attempt to lift the device, on 24 March, the lifting line was parted by chafing and the bomb fell to a new position on the seabed.

A further nine days elapsed before the weapon could again be found. *Alvin* discovered it 60 metres from the last known position and seventy-five metres deeper down the steep slope. The following day the *Alvin* again descended and once more attached sonar 'pingers'. At this stage the unmanned submersible CURV (Cable Controlled Underwater Recovery Vehicle), which we have already met in the submarine salvage chapter, was flown to Palomares from the United States.

So, on 5 April, after the fitting of an additional cable, CURV was guided to the bottomed weapon and managed to fix a good line on the weapon's attached parachute. Subsequent investigation showed that after this minimum essential disturbance, the weapon had slid down another 15 metres in depth. On April, CURV attached another line. And finally, on 7 April, CURV while attaching one more line, got herself entangled with the parachute. At which juncture, Commander Task Force 65 decided to take the responsibility of hoisting the whole *ensemble* upwards together.

As soon as the weapon had been hoisted up to the sort of depth workable by divers, the lifting arrangements were re-inforced and at long last the way was quite safe and clear for the bomb to be hauled up to the deck of USS *Petrel.* Thus ended a most unusual and most expensive operation which taught everyone so many valuable lessons; it had also entirely proved the competency of the United States Navy to discharge its responsibility to the other Armed Services.

Captain Bill Searle, a frequent visitor to Britain during his stint as US Navy Supervisor of Salvage, told me something about the excitements, frustrations and satisfactions obtained from the whole intricate operation. He also said that the Navy finally billed the Air Force for $5·5 million. That was no doubt to substantiate the old and true adage that: 'A good workman is worthy of his hire'.

There are, of course, other types of deep recovery which could only very loosely be connected with salvage in the strictest sense. Nevertheless, they demand mention as associated subjects. For instance, cable ships have, ever since the laying of the first cable, been able to grapple from the bottom and lift to the surface faulty sections of cable for repair and maintenance. Hydrographers have, since the beginning of hydrography, taken seabed samples in various ingenious ways. Offshore drilling rigs are also capable of working at great depths; as just one example Global Marines' *Gloman Challenger* worked in water down to 6,500 metres way back in the 1970s.

All these activities are cross-related and have important messages to pass from one to another. They all shelter under the omnibus new technology called Ocean Engineering. Not everyone recognises the existence of that new all-embracing technology but in other minds (including my own) it is the greatest enterprise of the future. And, the greatest hope for the world's ever-increasing population.

Probably the most significant deep sea recovery technique of the future is already being pointed out by the intermittent exploration of the ocean floors for metal nodules. It is quite evident that there is more good metal to be mined from the sea than from the land. If nothing else the area ratio of land to sea is a sufficient logical support for that concept i.e. 70 per cent of the earth's surface is covered by sea. Some of the metal does not even require 'mining'; it is there to be picked freely from the ocean floor by specially designed scoops.

Sea-water is self-evidently mainly composed of water and salt. But in smaller proportions, very many metal and metal compounds are present. Gold is the glittering but elusive prize and I recall an experiment which lasted for several years to extract that precious metal from sea water. The experiment ultimately failed because it was necessary to treat 3 million tonnes of water to obtain one ounce of gold. I think for the future, as far as can be foreseen, gold recovery can go to the bottom of the class. In very much more favourable circumstances, some of those seawater constituents coalesce round solid objects and turn naturally into nodules of useful metal compound. Many photographs have been shown in the newspapers and technical journals of selected seabed areas literally strewn thick with manganese nodules. There is no doubt that the chief factor adverse to recovery is the current high cost.

The costs, as estimated today, must however reduce in time by means of new underwater equipment, additional experience and good planning. That last is where the overall entity of Ocean Engineering comes in; advances and discoveries in any branch of it are bound to brush off on any other branch. In short, it is already feasible to recover metallic nodules in abundant quantities. But the economics of recovery must also take into account that vast new sources of supply from the ocean will inevitably lower the world's general overall prices. The whole concept stand or falls commercially on a complicated and continuing series of sums and only an eventual resolution of them will decide whether to go ahead. It is my view that the world will not be able to do without oceanic metals. In fact the real question is not whether we take them but just when.

Although cost is one more vital factor, the legal position is causing a bit of foot-dragging by 'doubting Thomas' types but certainly not by everyone. For instance, in November 1974, Deep Sea Ventures of America filed a claim with the United States Secretariat for an area of 60 million square metres of Pacific Ocean, some 5,000 miles offshore, and in water depths from 2,500 and 5,000 metres. Naturally the 'have-not' nations hotly dispute any such claims as 'big-stick' attempts to exclude them from sharing the resources of the deep oceans. Incidentally that area so claimed is an area specially rich in manganese nodules and other minerals are present in plenty.

The lawyers' arguments, not unnaturally, pertain to the time-honoured international standing of the 'high seas'; the 'have-nots' argue that although the high seas should be open to every vessel in the world for free passage those same high seas should certainly not be available to everyone in the world to grope about the bottom and pick up just what they want. The argument from those nations is that deep ocean resources should be run by specially organised and appointed international consortia for the good of mankind in general.

As a citizen of the world, I commend that view as a highminded and thoroughly civilised idea. But as a jaundiced realist, I reckon it will be the nations with the most advanced underwater technologies, and with the most warships to look after their own national activities, who will gather in the harvest 'fustest and mostest'.

I make no apology for apparent digression. This sort of deep sea recovery is definitely not salvage but equally truly it is so definitely an associated technique of recovery which may become greater than all the others. In the years ahead, when you consider that the seabeds of the world have sufficient metallic material to meet the *maximum* possible requirements of the whole world for twenty years – and very many years longer than that if demand is regulated carefully – then it is potentially the biggest recovery technique of all time.

It is estimated that there may be fairly easily recoverable amounts of nickel, copper, cobalt, iron, lead, aluminium in addition to manganese. It is further guessed that the Pacific Ocean alone had 1·7 million million tonnes of such dredgable materials. Add in all the other oceans and one can see what an important maritime industry is bursting to get started on full scale production.

In the absence of radical, new and universally agreed maritime law, it is well worth a re-furbishing of underwater techniques and of gun-boats for those talented and already rich, nations who have both those facilities at their disposal.

In the words of St. Matthew: 'Unto everyone that hath shall be given and he shall have abundance but from him that hath not shall be taken away even that which he hath'. How sad! How true!

On a final, and entirely unserious note, it will be extremely interesting to know exactly what else will come up from the ocean bottom alongside those valuable minerals. From time immemorial, seafarers have regarded the sea as their very own unfillable dustbin; the sea floor must be liberally spread with maritime oddments, large and small, which have been intentionally ditched or involuntarily lost overboard through the centuries.

Into the former category, fall the empty paint drums, jagged rusty wires and huge amounts of miscellaneous junk which hundreds of thousands of ships' officers (serving under every flag from the Phoenician to the newly emergent ones of today) have ordered to be consigned to the deep. In the latter category may be included a wide-brimmed Stetson hat which blew off my head in Chesapeake Bay and an expensive fountain pen which slipped out of my pocket when bending over a pilot ladder off, I forget which, the Canaries, Cape Verdes or the Azores.

Some future seabed exploiter may also discover the larger portion of a bridge telescope in mid-Indian Ocean. It got there when a sport-loving Second Mate bowled an old rubber door-stopper at me in fun as I took stance with the ship's telescope. I made a brilliant hook shot over my left shoulder but unfortunately the conclusion of the stroke found me only holding the 'inner tube' of the telescope – the remainder had flown off and was presently descending to Davy Jones's Locker. In passing, it may be said that this was a difficult incident to explain to the Master and to the Marine Superintendent – although I just fancy a merest flicker of amusement appeared on the face of the latter when the heat had gone out of the situation.

Chapter 11

Underwater work. Divers and submersibles

In the various recovery cases described so far, some idea of the several methods of underwater working has been touched on. But necessarily the description has been brief in passing only and confined solely to the actual job in hand. There is no doubt that a slightly expanded explanation at this stage will be helpful towards a better overall understanding of the subject. I will therefore go through the techniques – working from the shallowest inshore water to deepest offshore as a natural form of progression. I take it that the techniques of trawling and grabbing are so elementary and of such commonsense procedure as to need no amplification.

Diving

Few technologies can possibly have advanced so far and so fast as diving has done in the last few years. Techniques, physiology, safety measures, research, oxygen respiration, diving chambers with lock-in and lock-out facilities, helium mixtures, speech rectifiers, portable compression chambers, skin diving, seabed habitats in which divers live for weeks, better underwater illumination and so on almost *ad infinitum*.

The whole diving scene is an exciting, fascinating subject well worth a comprehensive book in its own right. But right here, we are concerned with search and recovery matters and must concentrate only on those diving operations which have a direct bearing on them. Everything else must ruthlessly go by the board.

Starting right ashore, at the beachline, we must first examine the functions of the swimmer-divers. They are most useful in the search phase of shallow-operations. Their most elementary search is the circular pattern which is nevertheless surprisingly useful if the position of the object to be recovered has been ascertained to a fair degree of accuracy. Briefly, a heavy sinker-weight is lowered as near to the estimated position of the object as possible. (Just twice in my long experience has the sinker been lowered bang on top of the object and thus effectively hidden it from searchers!). The diver then descends and secures his 'distance line' to the bottom end of the 'shot rope'; the top end going up to a marker buoy floating on the surface.

If the object to be searched for is large enough, the diver can swim a complete circle round the sinker at maximum length of his distance line – usually around a 15 metres radius. As the diver swims a complete circle, the object will snag the diver's distance line and thereby tell him to come inwards until he has located and identified the snagging.

For smaller objects, the procedure is for the diver to circle round the sinker, first on a very short piece of distance line looking and feeling as he goes round. If no result is obtained after 360 degrees of search, the diver lets out his distance line and repeats his circular search by an increase in radius equal to the distance he can see or feel. This goes on time and time again until he has found the object or has got to the maximum scope of his distance line.

If by that time the object is not located, that fairly large circular area can be marked on the chart as thoroughly searched but with negative results. The next step is to shift the buoy marker and sinker to the best alternative datum position and to go through the procedure again. The success of this particular method really depends on an accurately known first position because the second position of the marker buoy and sinker will do one of two things. Being of circular pattern the search from the second position will duplicate some of the searching from the first position. Alternatively if the two positions are more than a certain distance apart, the two circular patterns must inevitably leave unsearched areas of seabed between them – always a nagging worry to the oficer in charge of the search.

Far more useful for looking at large areas methodically in salvage-recovery is the so-called 'grid search'. In this method two wires are laid parallel on the seabed attached to sinkers on the bottom and marked by buoys at the surface. In between the two bottomed and parallel wires is a moveable cross-wire which can be shifted along the parallels bit by bit. Usually two divers swim, looking and feeling, crossways along the opposite sides of the cross-wire – thus sweeping quite a broad band. When one negative result crossing is completed, the cross-wire can be shifted along the parallels by slightly less than double the visibility distance or (i.e. in murky water) the 'feel' distance. By this method, a rectangular area can be very thoroughly searched and then the rectangular pattern can be moved very precisely to an adjoining position. By this means, large areas of the seabed can be searched with absolute confidence and certainty, albeit in a rather long time.

There are several other types of search but those two are the most common and the most useful ones in the line of business under discussion here. The finding of limpet, or other mines is an exciting procedure but fortunately none of our business. Thank Heaven! Let the cobbler stick to his last. Incidentally when a diver is looking for mines (an unlikely commercial pursuit) all equipment and fittings are non-magnetic so as not unduly to inflame the passions of magnetic mines.

For actual physical hard work and under-water cutting, welding, burning and other salvage tasks – the 'standard' suit is preferred by salvors to all lighter outfits. In the standard suit, the diver is completely (with the exception of the 'hands' when 'feeling') covered and thereby protected from abrasions. He is kept warmer, is in constant good telephonic contact with attendants on the surface and – by deflating his suit slightly – can 'anchor' himself on the bottom so as to use both hands for working. In addition the diver's head is well protected from accidental bumps by the traditional spherical copper helmet. That last item explains why these divers are distinguished from others by being known as 'hard hat' divers.

Without waxing too historical, or spending too much time on detailing the progressive improvements, it can be said that today's standard dress is not terribly different from the first type of enclosed dress sponsored by Augustus Siebe right back in 1837. As already said there is a rigid helmet, a double-texture proofed twill dress and a heavy vulcanised rubber collar to which the helmet is screwed. The diver wears warm underclothing, rubber rings are often fitted over the cuffs for extra watertightness, heavy boots are worn and heavy weights are attached to him – all of which further explains why the 'standard' diver can firmly 'anchor' himself in order to do hard and difficult manual tasks.

The standard diver is supplied by air through a hose from a reservoir in the diving vessel or boat which is kept topped up by a special air-filtering compressor. Although he is in contact by telephone, all divers and attendants learn a simple code of rope-pulling signals in case of emergencies. The diver weighs a great deal in air so that leaving and entering the diving boat calls for great care on the diver's own part and meticulous assistance from the attendants on their part.

When I have stood, fully booted and spurred on the rung of an iron ladder over the side of a diving boat, I have never failed to say a tiny but heartfelt prayer in my mind. It is that the attendants may screw my helmet on before the rung of the ladder gives way and sends me gurgling to the bottom without a helmet. But men who do the job every day of their lives never give that dire contingency a thought – at least, if they do, they keep pretty quiet. Stiff upper lip and all that lark.

In the diving scene, as in all walks of life, some exponents are better than others. And the best ones are worth pampering, encouraging, complimenting and generally treating as most precious possessions. In many types of recovery, the degree of intelligence, willingness and capability can expedite, delay or totally ruin a whole recovery project.

Experienced and good salvage divers, breathing air, can do well down to about 60 to 70 metres depth – and most will 'pinch' another metre or so on an important job without even being asked. But disregard of safe limits must *not* be encouraged. Air contains nitrogen and at great pressures, nitrogen becomes narcotic and so unbalances the diver mentally as to make him totally oblivious to his surroundings and to safety procedures. Deeper than 60-70 metres, other breathing mixtures – mainly that of oxyhelium – are used. And they bring in their train, other problems.

Helium under pressure does curious things to the vocal chords and, over the diver's telephones now come divers' distorted and incomprehensible voices similar to that well known cartoon character, Donald Duck. Voice 'unscramblers' are nowadays in use but even these electronic wonders do not provide a complete answer. The divers also feel cold and are mentally far from being at their best; this is especially apparent when trying to perform complicated jobs. For all these reasons – and I know that the dedicated divers will violently disagree with me – I personally believe that the diver ceases to the *best* recovery/repair/survey tool after a depth of 70 metres (plus) is reached. This will not always be so; nothing is more certain. But I sincerely and positively believe it to be the state-of-the-art as I write.

To be more explicit, the salvage/recovery requirement is not merely for a diver to descend to a great depth, sustain himself in safety on the bottom and to return to the surface. The requirement is to do useful and possibly hard work down there. For others – including the dedicated deep divers – that can do with saying again. The requirement is to carry out meaningful and practical tasks not just to go down deep and to come safely back. Phenomenal depths have been achieved in simulated deep dives in pressure chambers. They are significant as symbols of research progress. I salute the brave volunteers. But things *are* all different in a cold, wet, dark and hostile seabed environment.

One of the most beneficial of the recently developed techniques is that of 'Saturation Diving'. In the bad old days, a diver descended to the deep seabed, did a few minutes useful work and then had to spend a disproportionately long time decompressing himself on the way back to the surface. For the benefit of lay readers, the long decompression time was essential in order to liberate the compressed nitrogen bubbles in the diver's blood and body tissues very gently in order not to cause pain, severe cramp or even worse effects i.e. the well known diver's 'bends'. Complicated time and depth regulations had to be observed rigidly and the ratio of actual work time to total submerged time was resultantly poor.

The theory and practice of saturation diving has done more than anyone could reasonably have anticipated a few years back to improve that work to decompression ratio. The modern theory is that human beings can live without harm in a compressed atmosphere for prolonged periods and that it is therefore not necessary at all to return them to atmospheric pressure after every dive. The resultant practice is that divers are kept in pressurised 'living chambers' in reasonable shipboard comfort; they are conveyed to and from their underwater tasks in much smaller 'diving chambers' from which they 'lock-out' for operations and 'lock-in' again on completion.

In this manner, a very great deal more work can be put into any task, exactly when required, and divers can even be transferred from task to task inside their chambers. In due times divers do have to return to atmospheric pressure and the decompression process is then indisputably a very long-drawn-out affair. It is however, equally indisputably, far less than the sum total of what would be required if individual decompressions had to be carried out after every dive and this is the great net gain. It has the further advantage that the divers can come aboard and decompress comfortably at the most convenient operational time – say when the salvage ship is returning to harbour or steaming to another location. The divers are under full professional and medical supervision, visually and by telephone, throughout and have recreational facilities, albeit necessarily limited ones to pass away the time.

All the years I was connected with the salvage/diving business, I suffered one ever-present, nagging anxiety. What happens, I worried, if some mishap should occur to the ship while divers were locked in their chamber still decompressing.

On 15 August 1991, my nightmare came true. An oil barge running from a typhoon off Hong Kong was overwhelmed, capsized and sank in a very short time. Almost miraculously a great many survivors were picked up by rescue craft in awful weather but the divers locked in their decompression chambers had absolutely no chance. No slightest chance at all.

It was a tragedy, a fatal tragedy, scarcely bearable to think about. But divers are a tough brave lot. The money is good and the divers fully understand and accept the risks. There is no shadow of doubt that others will carry on this type of diving as if nothing had ever happened.

Habitats

For the benefit of readers to whom the subject of underwater habitats is unfamiliar, just one example will suffice to illustrate the practice. When writing the first edition of this book, I chose General Electric's *Tektite* project only because I met their engineers and scientists as early as 'Oceanology 1969' (an underwater convention at Brighton), was much impressed by their enthusiasm and (nothing like a personal compliment to engender interest and good will!) was presented with the special project lapel-badge. That Brighton occasion was the outset of their *Tektite 1* mission which was a fifty-eight day underwater stay at a depth of 15 metres. Since then the years have passed and other *Tektites* evolved. There is no point in changing the example now. Newer systems differ in detail, construction and sophistication but principles remain much the same.

The *Tektite 2* structure consisted of two vertical cylinders mounted on a rectangular base. Each cylinder has an upper and lower compartment. The lower compartment of one cylinder contains bunks and a small galley (kitchen to landlubbers). The upper compartment is a control centre furnished with operational and experimental equipment. From *that* upper compartment a tunnel runs to the upper compartment of the adjacent cylinder which contains most importantly the air conditioning and purification system. From there down to the lower compartment which has research and specialist analytical facilities. Down further from this section runs the trunk to the sea through which divers leave for, and return from, their missions.

The whole habitat is supplied with air, water and electricity from the surface support vessel. The equipment is designed to operate for as long as crew-endurance say thirty days, remains satisfactory and surface support can remain on the job. A crew change can be effected part-way through the operation if reason exists for prolonging it.

It can be seen that this sort of habitat is very well orientated to ocean floor exploration, marine biology, marine geology and oceanography. There have been many other experiments over the years by other firms and agencies with men living longer in underwater conditions and at increasing depths. They have, however, so far had little application to search and recovery. They could well one day come to play an important part in long-term recovery operations. But that day is not just yet.

So far as salvage and recovery potential is concerned, we can now begin to leave the diving scene. In these modern times, we have recovery divers ranging from the most primitive to the most advanced. In the former category the diving women of Korea and Japan certainly merit a brief mention; their elementary method is merely to jump overboard, suitably weighted and descend to a depth of 25 metres holding their breath. The average dive lasts thirty seconds of which only a scant fifteen seconds may be spent on the bottom; these forays were originally for pearls but are now entirely for the collection of sea-food. They have a rest period of thirty seconds and then repeat the operation; this means that each diver may clock up an astonishing sixty dives in each hour. They are, in their own simple way, as phenomenal as the most advanced saturation and habitat divers.

You may think that type of diving a bit of a digression. But it is not entirely so. Years back, I used exactly a similar technique in Port Pirie, South Australia, to locate and report the attitude of a lost anchor in 8 metres of water. Because of lack of experience, knowledge and maybe guts, I found that I could not shackle on a recovery line; it was, however, possible to direct a grappling procedure to do the trick quickly.

I don't know about the diving women of Korea and Japan but my ears did not come right for weeks. I wished then I had not been so full of enthusiasm and the spirit of one-upmanship when ideas were called for. The owners of the anchor made me a 'without prejudice' award of £5 – which in those far-off, hard-up days was some little consolation! The Port Pirie *Recorder* also published my photograph which was high fame indeed.

Submersibles

As in the case of new diving techniques, there are many new ideas coming along, many submersibles are now working all over the world. But again, the majority are not active in the sphere of salvage and recovery. Some are for surveying, some are for bottom sampling, for observation of marine life and fishing resources, for conveyance of divers, for pipeline operations, as tourist and passenger-carrying observation vehicles, a few small ones for pleasure purposes, some for rescue of personnel from sunken submarines and some can combine several of those functions. So once again it is essential to confine ourselves rigidly to those suitable for use in salvage and recovery.

Let us then first exclude the 'wet' submersible i.e. those which convey divers from task to task at ambient water temperature and ambient water pressure. It is obvious that these were very restricted in depth and in crew-endurance; in reality, they are only an aid to the diving capability discussed in the first section of this chapter. For recovery tasks, the suitable vehicles are then left in two broad groups – those manned in a pressure of one atmosphere and those which are unmanned and manoeuvred by remote control. Since, during the course of the book, several recovery operations have been described and the uses of the two groups fairly practically illustrated, it will now clarify the subject to explain the main details of one vehicle of each group.

In the original edition I wrote of the physical details and operational capabilities of *Pisces* and CURV as one worthwhile example of each group. In the current revision I considered deleting them and substituting mint-fresh modern examples. However, I have decided against that measure because *Pisces* and CURV are an essential part of the young history of submersibles.

Both have worked well and are still in use; they have achieved much distinction and richly deserve to be remembered as pioneers of new operational concepts underwater. Further, if a reader understands the simple examples, it will be easy to move on later to more sophisticated newcomers. Lastly, such is the explosive progress in the world of submersibles that anything now described as brand spanking new will be knocked off the top of the league before this book is printed and on sale!

So we now go to *Pisces* again. The series of *Pisces* craft from the first onwards are basically the same except that the capability in depth and other operational factors improved as the series progressed *Pisces I* had a maximum operating depth of 456 metres for instance and *Pisces V* could work at 2,000 metres.

They are all relatively unsophisticated submersibles compared with some United States and other craft. But the simpler one can get, compatible with efficiency and safety, the more economical the submersibles are in operation, – an evident commercial attraction. The *Pisces* series were constructed as two-sphere craft with one-sphere for the crew to lie prone in reasonable comfort so as to maintain a lookout, manoeuvre the craft and to use the external manipulative tools; the second sphere is the machinery space and is connected to the crew sphere by a battery compartment. Battery-operated motors drive the submersible at roughly 1·2 knots and she is steered by varying the speed of the two screws.

The craft ascends by the neat physical expedient of increasing her volume without altering the weight by pumping oil from the machinery sphere into inflatable bags; she descends by doing exactly the reverse procedure. Trim (i.e. attitude) is effected by pumping oil from the machinery sphere to subsidiary containers in the forward 'eyebrow' of the vessel. The submersibles are fitted with two echo-sounders, one looking down at the bottom and the other looking upwards at the surface; underwater telephone; very high frequency radio for use when surfaced; a tape-recorder for note-taking; exterior lights; a gyro compass; a sonar for locating or avoiding obstacles; underwater camera with video recorder. That sounds a lot but I say again it is comparatively unsophisticated as submersibles go nowadays.

So much for the technical details. But more importantly, I have seen the *Pisces*'s outside tools cut wire rope on the seabed, burn through steel plates and pick up pieces of wreckage with the manipulator claw. Add to that, the story of the salvage of the *Emerald Straits* already told and you can see why the *Pisces* boats have been so praised for economical underwater recovery. The vehicle was also particularly expert, after naval trial firings, at locating and picking up the valuable trial torpedoes which were fitted with 'pingers' to assist prompt location.

Now for the *un-manned* submersible and it is probably most useful of all to look at the old faithful which has had such great success – the CURV. Sadly this does not fall into my personal experience but its details have been given very frequently by the US Naval Civil Engineering Laboratory, whose concept it is, and they are therefore fairly well known.

CURV 3, as is to be expected, has greater capabilities than its two predecessors. To save back reference, it is as well to say again that CURV stands for Cable-controlled Underwater Recovery Vehicle. And it has proved itself most admirably in exactly that capacity on several important operations. It is far from handsome and, at first sight, hardly qualifies as a 'craft' as it is really no more than an assembly of equipment, sensors and instrumentation mounted on an open metal framework. An early model was capable of working to a maximum depth of 2,100 metres but even that considerable achievement has long been exceeded.

Being un-manned, it is self-evidently 'tethered' to the parent ship on the surface and gets all its orders from the control officer in the ship. It is steered to the target by adjusting the speeds of the four propulsion motors and has no conventional fins, rudders or planes. For the searching *role*, it is fitted with continuously operating active sonar, a passive sonar for listening, an altimeter (i.e. for above seabed distance, a depthometer (for below surface measurement) and a compass. For specific naval-type recovery tasks, it can also listen for and eventually locate any 'pinger' equipped weapon. In the final stages of the search, it inspects and identifies the located object by one of the two television cameras and it can also use a still camera for record purposes.

Even more important from the practical salvor's point of view is that CURV is fitted with a multi-function tool arm which can grasp items, cut metal, grapple and is adaptable at short notice for other purposes such as have already been described in the *Pisces III* and Palomares bomb incidents. As has been seen, it can be air-freighted from one part of the world to another at short notice and can operate from suitable ships other than its own parent vessel. There is no room for any possible British flag-wagging here. I consider the US Naval Civil Engineering Laboratory has a really fine vehicle to back for very many cases of very deep recovery.

The impassioned arguments by the pundits for manned

vehicles against unmanned vehicles goes ceaselessly on. And vice versa just as violently. It will doubtless continue in that fashion for many years ahead. All I can say is that there are 'horses for courses' and everyone wants to take the right horse to the right course. When Captain Bill Searle, US Navy Supervisor for Salvage, originally said there was no substitute for the 'Human Eyeball Mark I' on the seabed, I wholeheartedly agreed with him. And still do. But Bill, in response to a request from me in the last few months, kindly canvassed five leading underwater search specialists on my behalf. The concensus is that: The use of unmanned underwater vehicles, which is to say the use of a remote controlled camera (and eventually tools) is to be considered but an extension of the Mark I Human Eyeball Concept – when it is associated with a real-time control system which is, in turn, commanded by man'.

As in every sort of operation there will be special cases. There are some cases, for instance, when the recovery job is extremely dangerous by virtue of craft-attitude, by the position of the item to be recovered (say, hidden beneath a cliff overhang or on a steep seabed slope subject to mud avalanches and so forth). At such times the salvage officer in charge would want, if possible, to use an unmanned vehicle. He still has the unnerving knowledge that loss of the vehicle, or even damage to it, may halt the whole recovery operation until a replacement vehicle can be found. But one responsibility will slide off his shoulders; he will not be risking the lives of his men. And, thank Heavens, even in this material age, people are still more important than things.

Manned vehicles or unmanned vehicles? Again, in the words of the old adage: 'Yer pays yer money and yer takes yer choice'. That means the best choice available on each individual occasion – and every such occasion is hedged about with a different set of circumstances. So there can be no definite clear-cut decision for ever and ever. There is, however, another subject on which there most certainly is a definite conclusion. With the development of saturation diving, and ever more versatile submersibles, future years are going to see some recoveries from underwater which would be unthinkable even today.

Unthinkable that is, except the small band of dedicated underwater technologists to whom no sub-sea achievement is impossible. More power to their elbows. We shall see what we shall see and I guarantee it will be wonderful.

Chapter 12

Out-of-the-ordinary recoveries

So many cases occur of sunken objects – which have been submerged for many years and even sometimes for centuries – coming unexpectedly to the surface that it is difficult to know where to begin to spin this particular yarn. Perhaps the little things ought to come first. If this book were the size of an encyclopaedia the record could never be a full and complete one. So the incidents are related more or less in random fashion as they come to mind. It's easier for the author that way!

For instance, on 9th February 1961, the Palm Liner *Akasa Palm* had to have a World War I shell removed from her anchor in Liverpool. The shell had been down on the bed of Liverpool Bay for at least forty-three years and thousands of ships had during that time, anchored in the same spot. The *Akasa Palm* must have felt rather 'picked on' by a malign fate that she should be the one to pick it up.

Biblical scholars will remember well that Jonah was supposed to have been three days and nights in the belly of a whale. Many regard that story as purely allegorical. Yet, in February 1891, a very similar occurence was authenticated. While catching a big sperm whale off the Falkland Islands, the American whaler *Star of the East* lost seaman James

Bartley out of one of her boats. An immediate and diligent search found no trace of him and he had to be presumed lost.

The whale was finally caught and it was stripped of blubber all that day and all that night. By the following morning only the whale's bones and stomach remained. Miraculously, when the latter was cut open, the missing seaman was found lying unconscious. James Bartley eventually recovered physically if not entirely mentally. It was said, however, that his skin was permanently marked and wrinkled by the whale's strong digestive juices.

Another extremely strange case was that of Lieutenant Michael Fitton and the *Ferret*, tender to HMS *Abergavenny*. The lieutenant was fishing for sharks off San Domingo and his pastime had an exceedingly curious outcome. In the stomach of one shark he found a bundle of papers which had been deliberately thrown overboard by the American brig *Nancy* when intercepted by one of the patrolling Royal Navy ships on 28 August 1799. The false identity of the *Nancy* and the false story she had told were thus proved beyond doubt in court at Port Royal. The ship with cargo was therefore confiscated by the British on 27 November 1799. An amazing bit of luck – good or bad according to which sides the parties belonged!

In 1830, HMS *Thetis* was carrying £160,000 in coin and bullion; she ran hard aground and sank at the foot of steep cliffs at Cape Frio Island off Brazil. Captain Dickenson immediately had the idea of salvaging the bullion and even more commendably actually improvised the means to do so. Diving bells were made out of metal water tanks and hoses were constructed to convey air down to them from the surface. He lashed together masts and spars from *Thetis* to make a heavy lifting derrick which, from a nearby cliff, raised and lowered the diving bell.

Dickenson suffered many setbacks from weather, sickness and failure of improvised gear in addition to dealing with unwanted 'helpful' advice from senior officers far distant from the scene. Later, when the improvised derrick finally gave way, he ran an overhead cable from cliff-top to cliff-top and used tackle suspended from that for raising and lowering the diving bell. After fourteen months work with an enthusiastic crew, some £120,000 of cargo had been retrieved. He was then ordered home but not before he gave a good turnover to his relief who brought up another £32,000. Captain Dickenson was a natural salvage officer bursting with ideas and practicalities. He got little credit in his own day and I am glad here to put his excellent work on record for posterity.

Less valuable but equally remarkable was the recovery of a large cargo of whisky from the barque *Scottish Prince*. This vessel sank off Southport Beach, Brisbane, in 1887, and the recovery was made in 1958. The whisky was still quite drinkable and – despite a slightly strange *bouquet* – still enjoyable.

Talking about whisky, I remember a much less happy story about a coastal steamer which sank at her berth in the River Tyne. Some time later, the vessel was successfully refloated and the various contents of the various compartments were checked as the vessel became available for inspection. That was a bad day indeed for the former master of the ship when a goodly number of bottles of spirits and an amount of tobacco was found in his cabin. He had neglected the essential little formality of declaring them to HM Customs officers on entering port. This particular salvage therefore cost him a considerable fine after prosecution in Court – and the confiscation of the goods. If anything can be a consolation to him, there is an old and apt precept that says: 'No one can win 'em all!'.

In Australia, a seemingly unindentifiable ship in 15 metres of water off the Queensland coast was ultimately identified after a good bit of detective work. An old safe was brought up and a still-existent firm of safe-makers in London solved another mystery by stating to which ship it was originally consigned. It belonged to the Adelaide Steamship Company's *Yongola* which had disappeared after leaving Mackay, Queensland, in March 1911.

The entry of 'amateurs', i.e. sub-aqua clubs into the

underwater business has meant the finding of a great many long-forgotten old wrecks. Wherever the sub-aqua clubs go which, especially in the shallow waters surrounding the British Isles, is pretty far ranging, they discover much of archaeological value and occasionally of financial value. Marine archaeology is, of course, a subject of its own and is hardly a recovery job except as regards recovery of knowledge and artifacts or coins.

It is of interest in passing that the bad old days when just anyone could descend on and plunder just any wreck have gone. Much of maritime archaeological value has been ruined for ever by ignorant, greedy or thoughtless pillaging. But nowadays, around the United Kingdom, permission has to be obtained to work on historical wrecks and strict rules have to be observed. This is a real advance on the careless, reckless, wasteful diving work that used to go on only a few years ago.

'Most modern archaeological projects are concerned with accurate marking, sketching, photography of the ship-remains and recovery of a few selected items. The 1545 sinking of the *Mary Rose* in 13 metres depth in the Solent is a case in point. The wreck, which had first been located in 1936 had to be laboriously relocated in the 1960s by sub-mud surveys, and top-class sonar equipment – at which juncture the *Mary Rose* (1967) Committee was formed to regularise and safeguard the legal position.

So far the vessel has been accurately surveyed and plans of the complete vessel have been drawn up from the practical surveys augmented by theoretical naval historical knowledge. Further the ship's attitude and condition is now well known. Several items, such as gun barrels and tableware have been recovered are are being carefully restored. There is, more-over, just an outside fighting chance that half the vessel can be raised intact. The money to do it is – as always – the make-or-break aspect'.

Since I wrote that in 1977, a dedicated force of enthusiasts drove on through difficulties of weather, tides, techniques and finances to an astonishing success. Much of the structure of the ship was recovered and preserved together with many real archaeological and historic treasures. The results can be viewed daily by visitors to the Royal Naval Museum, HM Naval Base, Portsmouth. A full account is available in books which I shall list in the bibliography at the end; one is by Doctor Margaret Rule who got her teeth into the project like a terrier and never let go. The other by master-diver Alexander McKee.

One piece of marine archaeology which did definitely fall into the salvage/recovery scene was the magnificent Swedish feat of raising the *Vasa*. That vessel was originally the pride of the Swedish nation, a then great naval power, it capsized in a squall on 10 August 1628. Thereafter she rested on the bottom until she finally broke surface again on 24 April 1961; it was a 'proper' and marvellously executed salvage job which is worth describing in some detail.

The Swedes were fortunate in having on hand a national marine salvage firm, the Neptun Salvage Company which ranked as one of the best in the world and had done so for well over a century. The raising of the *Vasa* put the firm's ingenuity improvisation, expertise and practical work to probably its greatest test ever and it succeeded brilliantly. I think it well worth saying at this stage – and every professional salvor will, from personal experience, understand perfectly – the Neptun Company did the major portion of both planning and practical work on the project. Yet there were quite a few much less involved organisations which utilised the operation in support of their own public relations efforts! Alas! 'Twas ever thus!

I name no names, ever bearing in mind that evergreen seamanlike precept: 'No names, no mastheading'. I am too senior to be mastheaded or even keel-hauled but nevertheless no more at all on that score. If anyone reads the foregoing and the cap fits, will they kindly put it on and wear it.

The salvors were smiled on by Lady Luck in just one or two respects. The vessel had gone down quickly in splendid condition, with excellent equipment and fittings, in a comparatively quiet and clean area of sea. (In passing, it is of interest that in contrast to the *Vasa*, the timber of a Roman vessel discovered in the Thames in 1910 was reduced to the consistency of mushroom). Subsequently, *Vasa* became covered in mud which acted as a further protective factor. Nevertheless, all that having been said, the complete recovery of her after 333 years of immersion was a really wonderful achievement.

In brief, the operations went in almost the customary, classic, stage-by-stage salvage progression. First came the location which was a longish task. Then came the divers' detailed examination and reports from which resultantly a plan was conceived by Neptun's Salvage Superintendent, Captain Axel Hedberg, to lift the wreck in one piece. Tunnels were dredged under the ship and six pairs of heavy wires were passed through six tunnels. Lifting craft were brought in, 'pinned down' and a lift carried out. This pinning down and lifting process was repeated eighteen times to shift the *Vasa* very gently indeed from about 35 metres depth to a more favourable position in a depth of 17 metres so that work could proceed more easily.

Thus ended the first stage of the operation – which had lasted from 20 August until 16 September 1959. A pause was made at this point for several reasons; because of the onset of the winter season, for rethinking, for brushing up the planning and for advancing ideas about *Vasa*'s final programme after the actual salvage had been completed. Not least in the thinking were the legal financial and operational responsibilities of the various contributors.

Stage two began in 1960 with clearing the upper deck and upper gun deck so that they could be strengthened to withstand the pressure of the pontoons. The hull openings (and there were very many) were sealed by divers before further operations began. On 20 March 1961, the pontoons arrived back on the job. The lifting wires were repositioned to give good and exactly equally distributed support to the *Vasa*'s approximately estimated 400 tonnes weight. Lifting operations were repeated and on 24 April the old warship was at last beginning to show herself, bit by bit, on the surface.

At the appropriate juncture, powerful pumps were placed

on board and emptying of the hull commenced. By 4 May, the vessel was high enough in the water to pass over the sill of the selected drydock at Beckholmen even allowing for the slight list she had still retained. She was docked down on a huge, specially constructed, concrete pontoon which was to be the *Vasa*'s future permanent floating home. The vessel was propped up and shored in selected places to prevent sagging and bulging through any structural weaknesses which had developed during the 333 years immersion.

That ends the pure salvage story. But for the sake of completeness, it should be emphasised that it was the beginning of the fascinating archaeological story. Over 20,000 objects in total were carefully removed from the *Vasa* for careful examination and, where possible, for preservation. Mud and ballast was discharged as quickly as practicable to lessen the 'in air' stresses on the ship. Although the wood was in mainly good condition, the exterior surfaces had to be chemically treated and the ship had to be covered and kept moist by means of water sprinklers. In other words, the ship, after such long immersion, was allowed to dry out and adjust to above-water conditions in a slow and carefully controlled manner. The fact that the vessel is now a floating museum with exhibition, lecture and film halls as well as offices and maintenance shops is proof of the success of the post-salvage part of the project.

The whole *Vasa* project demands great commendation – for the original visionaries, the planners, the archaeologists, the scientists, the sponsors and, most of all, the salvors. Because this is a salvage book, I must concentrate on that last category. In all the long history of the Neptun Company there can never have been a more complicated or a more successful job. Immense credit goes to Neptun for ready advice, steadfast resistance to that 'helpful' interference *always* present at every big salvage job and great courage in pressing their ideas into successful reality.

Salvage Superintendent, Captain Axel Hedberg, must qualify for the major share of the compliments because of keeping his plans and his confidence going despite the critical eyes of millions of people on him – not only those of the now extremely interested general public but also those of all his professional colleagues in every maritime nation. But the whole company and every employee engaged on the job really deserves some goodly part of the commendation. Every part of the salvage service was given to the project entirely free except where the company incurred genuine expenditure over and above their normal running expenses. In those cases, and in those cases only, they claimed and received 'out-of-pocket' payment. This generosity was a grand gesture coming from a commercial company normally in business for the entire and understandable purpose of making a profit.

Now back from the ship-salvage to the artifact recoveries. In 1798, HMS *Colossus* sank off the Scillies. She was on her way home with many cases of valuable pottery on board; the contents of the cases were the personal belongings of Sir William Hamilton (husband of Nelson's Emma) which were being returned home separately after William's ambassadorship at Naples. The vessel has been located and much of interest and value has been recovered by Mr Roland Morris, and his team of divers, in remarkably good condition. This project has proved of a most useful nature to historians, museum experts and naval architects.

Not all such finds are acclaimed unanimously by all the experts. For instance I noted way back on 12 February 1957, that *The Times* (off caps and stand to attention!) said that the anchor of Bligh's *Bounty* had been found and recovered after 167 years of being submerged; the report went on to state that: 'the anchor with 7 feet of chain attached, was raised by a group led by a Seventh Day Adventist pastor, Mr Leslie Hawk, and an American yachtsman, Captain Johnson of the brigantine *Yankee*'.

Three days later a letter to the Editor appeared in the same newspaper from Mr Alan More, saying: 'The mutiny took place in 1790 but chain cables were introduced into the Royal Navy in 1811'. Therefore, suggested Mr More, the anchor could not credibly have belonged to *Bounty*. As a writer myself, of long and hard experience, I know that there is always at least one alert and eagle-eyed reader like Mr More who is minutely scrutinising my contribution. And although contradictions are a considerable blow to one's professional ego, it is impossible not to welcome the criticisms as advancements in knowledge. Nevertheless, I still hope against all odds not to get too many such advancements in knowledge from readers of this book!

As for treasure and treasure ships I intend not to discuss the subject in detail. That treasure has been recovered in the past is indisputable. And that treasure will be recovered in the future is equally certain. But looking for treasure is absolutely no way of *ensuring* oneself a financially independent old age. But hope (as they say) springs eternal.

There are few parts of the world without their own treasure ships. The Dutch *Gilt Dragon*, sunk in 1656, near Fremantle, Western Australia. The Vigo Bay galleons of Spain's 'silver fleet'. The *General Grant* sunk in 1866, off the rocky western shore of Main Island, New Zealand. The *Royal Charter* lost in a gale off the Anglesey coast in 1859. The *Grosvenor*, an East Indiaman, which struck submerged rocks 15 miles north of Port St John, South Africa, in August 1782. The *Birkenhead*, off Danger Point, South Africa, a casualty of February 1852. The *Lutine* wrecked in the shallows between Terschelling and Vlieland in October 1799. The galleons of the Spanish Armada, of which the *Florencia* (the Tobermory galleon) is the most famous and the *Santa Cruz* in Cardigan Bay is the runner-up.

One could go on cataloguing treasure ships almost for ever. There is no doubt that some of these vessels did carry treasure and therefore searchers have had, and will have, lucky strikes. But what astonishes me is that so many wrecked ships just happened to be carrying such large quantities of gold and silver. It does appear as if no ships through the ages ever did go down carrying an ordinary, common-or-garden cargo or even in ballast! Curious to say the least.

In this connection, there was the salvage of the *Merida* just before World War II. She had sunk in 1911 off the coast of Mexico with (as always!) vast quantities of gold and silver plus, for good measure, the Crown Jewels of Mexico. After many years of search and finally eight months of very hard work, a salvage team managed to cut its way through to the strongroom and recovered the ship's safe. Sadly for them, it was unlocked and quite empty – indicating that either someone had been there first or that the crew had not abandoned the ship without attending first to their own personal prosperity.

Again, in 1937, Captain Simon Lake found the 'treasure ship' HMS *Hussar* which had sunk in the East River, New York with the customary treasure on board. In that ship, the salvors sadly found American coins only to the value of approximately forty-one pence. As a final note on a similar subject, a man was decared bankrupt at Hastings Bankruptcy Court in 1957. He had spent £4000 of £23,000 inheritance (a good solid amount of money at that date) on setting up an expedition to find Captain Kidd's buried treasure. He had the map!

If you begin to think that I believe searching for treasure as a professional task to be a fairly insecure occupation, you would be dead right. In my opinion it's better, say, to be a bus conductor paid time-and-a-half on Saturdays and double-time on Sundays. For every expedition which has been successful, there have been hundreds of unsuccessful ones.

But don't let us end on an entirely gloomy note. Even in the short period of 13 years, since I wrote my last comments, the scene has changed dramatically for one class of treasure searcher – the expert, well organised, well researched, well financed expedition. But not at all for the enthusiastic amateur with a small boat and a big hunch.

How come, after so many centuries of trying, a few short years can transform the outlook? First there has been a great improvement in the quality of pre-operational planning and better detective work concerning the vessel in question, i.e. more accurate position of sinking, and the whereabouts of valuable items within the vessel itself. Thus the initial, and always expensive, general area of search is minimised.

Next, the acoustic search tools have improved sensationally.

Wonderful seabed pictures are obtainable by side-scan sonar from a surface vessel in reasonable depths and by towed, well submerged, instruments in deeper cases. Wreck identification has been made easier by such pictures augmented by improvements in diving and submersible capabilities.

But the greatest bonus of modern search operations is the current astonishing accuracy of position fixing. In former days astronomical observations (and only then when weather was clear enough to take them) fixed the ship's position within one mile at best. Early Loran improved the accuracy to half a mile and sometimes even a quarter. Since that, Loran has again improved but the real wizard is the new differential Global Positioning System using satellites and I have seen claims of plus or minus 3-4 metres which is equivalent to two tall men laid end to end!

No surface vessel is able reliably to navigate to that degree of precision but one can see the path which has been left behind and thus be assured that no gaps have been left in a search pattern. And 'targets' once identified as not required will never be accidentally worked on again. All those factors are important in saving time. But the daddy of them all is the ability to return to the right contact time and time and time again without difficulty or doubt. In former days, a support ship leaving her station for safety or logistic reasons came back and started the location/identification process all over again.

I am just going, as briefly and non-technically as possible, to tell a very thrilling story I was initially undecided whether it ought to go in the 'deep recovery' chapter but decided on this place because we have just been discussing the viability of treasure hunts.

Way back on 12th September 1857, the paddle steamer *Central America* sank in violent weather about 160 miles off the coast of South Carolina. She was carrying 476 passengers, 102 crew and no less than three tons of gold. The amount was known accurately because it had been transferred from another ship and tallied carefully by both parties. The reason for this great amount is that a routine bi-weekly shipment from California to New York was augmented by the amount carried by passengers returning from the Californian goldfields.

In turn, the reason for this indirect route was that from California to Panama by ship, land transport across the Isthmus and by sea again to New York was the quickest and most comfortable one available. Coast to coast railway systems (and the Panama Canal) were still in the future. Incidentally a contemporary newspaper account said that 153 persons had been rescued and that the remainder had perished. So much for history. The whole thing is a wonderfully gripping story but this specialist book is no place to be telling it.

In 1977, enter Thomas G. Thompson of Columbus, Ohio. He was an ocean engineer with a hobby of studying shipwrecks in general and *Central America* in particular. In 1985, after a very long feasibility study, Thompson founded the Columbus-America Discovery Group (CADG) with two former school friends and the project was underway. The feasibility study was so careful, so comprehensive and so honest about chances of success that a near miracle happened. When approached, 106 hard headed businessmen in his home town (most of whom would not normally have invested in a treasure hunt in a thousand years) raised seven million dollars! And later, when Thompson required more money, the same investors put another $5·1 million in without demur. Some stake. Some confidence!

With the money now available, CADG bought a suitable vessel (a small Canadian former icebreaker) manned her, fitted her out, stored her and equipped her as necessary with search and recovery gear. That last category included magnetometers, ground penetrating radar (GPR), sub-bottom sonar profilers and wide-swathe sonar – all of which were linked up to display real-time images and manual control in the operations room. In 1986 she surveyed 1400 square miles of seabed and, by September, the wreck was considered to be found. The search costs ran at $20,000 a day. It can be seen why 'Time is of the essence' in recovery operations.

In the meantime, the Remotely Controlled Vehicle (RCV)

Nemo was being planned, designed and built. Its great advantage was in being purpose-built for one specific task and not just a vehicle borrowed and adapted as requisite. I have not been able to study a sufficiently detailed sketch (maybe because it is commercially confidential?) to describe it here. But it was said to have instruments/tools operated from the surface vessel in real time and capable of picking up objects as small as a single coin or as large as 300 tons.

In summer 1988, CADG marked out a very accurate grid, a kilometre square, with sonar beacons at each corner. This was more for the benefit of archaeologists than for cargo recovery. And *Nemo* recovered the ship's bell as certain proof of identification. In 1989 the full team arrived on site and, by October, had retrieved over one ton of gold from a depth of 8000 feet.

Since that time, recovery has gone on apace. The final weight has not been announced. By the same token, the *value* cannot possibly be known for a good many years ahead. So much of the recovered material is in actual gold and silver coinage that it is a numismatist's dream and much enhances the 'raw' value of precious material. Released on to the market skilfully, never putting forward too much at a time, the kitty will get more and more valuable as the years roll by.

One plausible report is that if all goes well, each of the initial investors will be paid back 85 times the original investment. In the words of the old song: 'Nice work if you can get it. So get it if you can'.

If all goes well? But now there is a sour note. CADG's attorney had to go to Court to obtain injunctions against two lots of salvage competitors who tried to play on the same pitch. Then, when the gold started to come ashore, no less than 39 insurance companies said that they had paid out on claims in 1857 and therefore owned part of the recovered gold. Plus the University of Columbia and three other persons who said that they had assisted Thompson. But they all failed to convince Judge Kellam in the District Court for the Eastern District of Virginia (Norfolk Division).

The Judge dismissed the insurance claims because of lack

of evidence and the fact that none of them had ever made any efforts at recovery in the elapsed 133 years. The other claims he dismissed because he considered that no claimant had given any useful assistance. Of course, there are higher Courts for appeal and the matter is not yet settled. At very least CADG will get its investment back with handsome awards for its great and successful efforts, At best CADG will scoop the legal jackpot and I hope that they do.

Chapter 13

Marine insurance. Legalities

The saving of money is the most important single reason for setting any commercial salvage project into operation. That's for sure. So no book on salvage can possibly go without some reference to those people who have the most intense financial interests – the underwriters.

The principles of insurance go back almost as far as human history has been recorded. The real historical enthusiasts aver that one of the first recorded insurances was when a certain Pharaoh of Egypt and his adviser Joseph carefully stored grain in the years of plenty to guard against the lean years which were statistically bound to come. Be that as it may, that was non-marine insurance; perhaps the earliest known marine insurance was back around 3000 BC when Chinese traders on the Yangtze River distributed their cargoes over a number of vessels so as to average out the losses amongst themselves.

Of fraudulent claiming, the first evidence comes from the pen of Livy in about 215 BC when the State was, by necessity underwriting the carriage of war material by sea: He wrote:

> Because there was danger to the public from the violence of the weather and the transport of stores to the armies, the farmers of the customs would make false reports of

shipwrecks and the very shipwrecks which really did take place and were truly reported were occasioned by their own fraud and not by casualty. They would put a few things of trifling value on board old and shattered ships and when they had sunk those ships in the sea, the sailors would escape in boats, specially prepared for the occasion and then falsely pretend that a great deal of merchandise was on board!

Old customs die hard. Again I will name no names but similar attempted deceptions have happened within the last decade and they always will when there is any chance of easy money to make by the inscrupulous operator. *Plus ça change plus c'est la même chose*. No wonder the underwriters have the reputation of being ultra-careful before paying out hard cash!

Once more, on the principle of writing about what one knows best. I select Lloyd's of London as the body of underwriters to comment on. Fine marine insurance facilities exist in other countries but most British ships and a great number of foreign ships are insured at Lloyds – and the vast majority of the casualties with which I have been associated have been so insured. Quite naturally, if the risk is great, say in the case of monster tankers, then the various centres of insurance re-insure with each other to minimise and spread the risk. This is no book to load with statistics but as one random example, the 1976 payout on the Norwegian tanker *Berga Instra* was $18·2 million and about 70 per cent was covered in London. Truly marine insurance is one of the most international of businesses.

A brief description of the Lloyd's organisation will therefore assist the laymen at this stage. As almost everybody knows, it all started in the seventeenth century when a Mr. Edward Lloyd ran a coffee house. There were many such coffee houses. The 'America and New England' was frequented by commanders of ships for North America. The 'Amsterdam' had a clientele of employees of the Hudson's Bay Company. The 'Coal Exchange' was where captains of colliers met to arrange for carriage and sale of coal freights.

The 'Jamaica' where it was said: 'one sees nothing but acquatic captains in the trade of that island'. Then there was 'Sam's' where reports stated that 'the commanders to be spoke every morning'. Mr. Edward Lloyd's establishment happened to be one where the customers' main business was marine insurance and the reputation of the meeting place increased as time went on.

The present-day Lloyd's, incorporated by Act of Parliament in 1871, runs on the same principle. It is not an insurance organisation in its own right but merely provides the facilities, accommodation and correct setting for the underwriters (or frequently, syndicates of underwriters) to pursue their individual businesses just as Edward Lloyd provided them in his day. The Corporation, however, has the strictest rules about the integrity and financial status of members, the provision of adequate security, the payment of premiums income into a Premium Trust Fund, the payment of a levy on premium income into a central fund to protect insurers against any member not being able to meet his commitment. All this, together with the annual audit which all underwriters must submit makes the business conducted at Lloyd's to be regarded throughout the world as model of reliability despite current shipping.

A shipowner, seeking insurance, places the matter in the hands of a Lloyd's Broker who thenceforward becomes the representative of the owner and not that of the underwriters. The broker will have a shrewd idea of the going rates and will shop around the underwriters for the best bargain he can get for his client. So much varied business is done by underwriters that no one can be expected to be an expert in every subject. The broker will therefore select an underwriter who is an acknowledged expert on the specific matter in hand. This chosen underwriter will assess the risk, quote the rate and – if desired – accept a portion of the risk by initialling a slip. This 'slip' the broker is now able to take around other underwriters (who will accept the judgment of the expert or 'leader' for further part acceptance until the slip is completed for the total sum required.

Once again, the confidence between, underwriters and brokers is quite unique. Although in no way a legal document in itself, the broker knows that anything initialled on the slip will be strictly honoured even before the policy is formally issued.

Because of the immense financial risks involved, the underwriters – as said at the beginning – are quite the most interested parties in the outcome of any commercial salvage case. They will frequently instruct the Salvage Association (also headquartered in Lloyd's building) as a body skilled in such matters to look after their interests. The motto of the Association is *Quaerite Vera* or 'Seek the Truth'.

The Association's activities are controlled from London but it has numerous branches throughout the world. It is capable of providing expert knowledge, assistance and advice to minimise loss or damage to ships and cargoes. It operates no salvage equipment and no salvage ships and is entirely non-profit making.

The master of a ship in trouble will usually – although underwriter's and owners' interests do not march absolutely hand-in-hand – do well to accept the advice of the Association's salvage officers. No salvage officer, however illustrious, can usurp the Master's unique responsibilities for command of the vessel, but compromise and co-operation can go a long way towards a successful operation. In passing it, should be said that the services of the Salvage Association are available to other reputable organisations besides the underwriters. Anyone who is seeking the truth about salvage is in the same line of business!

However the arrival of an Association (or any other) salvage officer must inevitably take some little time and the shipmaster cannot fail to have some worrying time in the early stages. And, in this connection, one of the finest things that ever happened for practical men involved in a marine salvage crisis was the institution, in 1910, of Lloyd's Standard form of Salvage Agreement. Widely and familiarly known as 'Lloyd's Open Form', it has become so trusted on the seven seas that almost any Master of a vessel in trouble (and his owners) knows that by accepting its terms they will get a fair deal from the Arbitrator appointed by the Committee of Lloyd's.

The form permits the men at sea, possibly even dicing with death and disaster, to get on uninterruptedly with the job in hand in the sure knowledge that every legal and financial detail will be sorted out in a calmer atmosphere ashore. It also means that expensive litigation in the courts is done away with which is particularly useful if the court of the country in which litigation would take place has had little experience of such matters.

The 'Open Form' contract is a NO CURE—NO PAY arrangement. In other words success, or partial success at least, is the only way the salvor becomes entitled to payment. The most hard, prolonged, brilliantly executed operation which for some reason ends in failure is entitled to exactly nothing. Briefly the salvor may make reasonable use of the ship's gear free of expense. As soon as the services are completed (and in practice, usually beforehand) the salvor notifies the Committee of Lloyd's as to how much financial security he requires to be deposited and until that security has been so deposited with Lloyd's, the salvor has a maritime lien on the salved vessel and cargo; she must not be removed from her berth without the salvor's consent.

The owners of the salved property may accept the correctness of the initial deposit assessment without demur as reasonable. Much more usually, however, they will ask the Committee of Lloyd's to appoint an Arbitrator; the arbitration to take place in London according to English law. In most cases the Arbitrator's ruling is accepted by both parties but there remains a right or appeal to a further Arbitrator should one party or other feel sufficiently aggrieved.

Lloyd's 'Open Form' has to be signed, before work commences, by the salvage officer or Master of the salvage vessel on one hand and the Master of the casualty on the other as the representative of his owners. On one occasion, however, I recollect that weather conditions made inter-ship contact impossible and the verbal question (Do you accept

Lloyd's Open Form?) was put to the casualty by loudhailer in full hearing of the crews of both ships as witnesses. On receiving assent by similar means, work started immediately and the form was signed at the first subsequent opportunity.

The form is most certainly an enormous improvement over the 'bad old days' when the would-be salvor explained to the Master of the casualty with persuasiveness and occasional bullying what desperate danger he was in – especially if the salvage vessel decided not to assist and went off to leave him to his fate. Thus, in more than a few cases, exorbitant promises of reward were exorted from Masters in dire anxiety.

The Arbitrators are skilled lawyers well versed in the practices and traditions of the sea – and in reconciling the contradictory viewpoints of the parties concerned. The Master of the casualty will frequently advance the point of view that his vessel was in minimal danger; the salvage vessel, on the other hand, is bound to express an opinion that the casualty was in maximum danger and that some dire disaster was imminent had he not stepped in and averted it with consummate skill! It is one of the great wonders of the world how differently two experienced seamen working within a few metres of each other can view the same situation!

I recollect a case of a vessel, hard aground, on which the salvage officer, salvage vessel and salvage team spent weeks in getting the vessel to exactly the right condition for refloating. All preparations being at last complete, the salvage officer ordered all concerned to get a good night's sleep in view of a final extra hard day's work ahead. But one man had different ideas. The Marine superintendent of the company owning the stranded vessel called his own crew out early, put the vessel's engine to full astern and hauled the salvage wires successfully to bring the ship afloat. Alas, however, his subsequent contention that his crew got the ship off unaided by the salvors was not looked on with favour by the Arbitrator!

The size of the salvage award must inevitably be related to the value of the property saved. To take the point of absurdity it is no use a salvor spending £0·75 million on the most magnificently successful operation of saving a £0·5 million ship and then to expect any profit out of the job. On the other hand no Court and no Arbitrator will be mean to a dedicated salvor. One particularly good reason is that salvage of ships and property is quite evidently desirable to underwriters and any award must be generous enough to encourage other salvors to make similar effective exertions in similar circumstances elsewhere, either in territorial waters or on the high seas.

As I first sat at my desk at the Navy's Salvage School half a lifetime back, I was taught the various constituent elements going into the assessment of a salvage award. I thought my instructor wise and experienced beyond all others as I wrote them down. Much later I discovered he had cribbed them from the book of Sir W.R. Kennedy, the greatest British (and probably the acknowledged world) – authority! They have not changed in the last quarter century and I quote them as then noted.

(*a*) As regards the salved property:
- (i) The degree of danger, if any, to human life.
- (ii) The degrees of danger to the property.
- (iii) The value of the property as salved.

(*b*) As regards the salvors.
- (i) The degree danger, if any, to human life.
- (ii) The salvors classification, skill and conduct.
- (iii) The degree of danger to the property.
- (iv) The time occupied and work done in the performance of the salvage service.
- (v) Responsibilities incurred in the performance of the salvage services e.g. risk to insurance and liability to passengers and freighters through deviation or delay.
- (vi) Loss or expense incurred in the performance of the salvage service such as e.g. detention, loss of profitable trade, repair of damage caused to the ship, boats or gear, fuel consumed, etc.

As regards the conduct of operations, no Court or Arbitrator expects perfection. They recognise probably more than anybody else that: ‘Only Allah is perfect’. So long as good planning, good seamanship and good commonsense is used, they will be satisfied.

It is only to be expected that learned counsel for the owners of the casualty will suggest to the salvage officer that had he done something different, the result would have been very much better! That is his job. It brings to my mind just one more attribute of a good salvage officer i.e. after all the stress, strain, physical hard work and general exertions of a salvage job, he must remain patient, even-tempered and coherent when under heavy implied criticism in a cross-examination probably many months later.

Out of the experience of years of backseat driving from a desk ashore, I have only three pieces of advice to the salvage officer. I am sure the lay reader will agree with them entirely.

First, be completely honest. Remember the advice of that outstanding United States Admiral, David McDonald; ‘Always tell the truth and you never have to remember what you said’.

Second, no matter what the pressing emergencies of the salvage situation, never fail to keep an hour-by-hour (occasionally minute-by-minute) detailed narrative going. Times, logistics, thoughts, weather forecasts, discussions with others (most particularly if you have had to reject, for good reasons, advice or suggestions offered by others) and so on. Time thus spent is invaluable in recalling, sometimes after many months have passed, why you did something in preference to something else. In these days of miniaturised tape recorders there is absolutely no excuse for failing to keep a full record.

Third as soon as the operation has finished, and while the memory is still fresh, write a methodical, well-ordered, Salvage Report in draft form. Get a fellow salvage officer or any office administrator to examine it closely. He or they may think that there are important omissions, that one or two points might benefit by addition of further detail, that the wording might be tidied up to make certain sections more readily intelligible or that a point or two might be presented in a different order. If you agree, then re-write the Report and have it typed. Sign it and stick to it through thick and thin. Reject any suggestions to 'cook the books' by omission of something someone finds better not said nor even by placing extra emphasis on some point considered favourable.

The report is a permanent record of the Salvage Officer's own integrity and reputation which will be at stake in the Courts or in front of the Arbitrator. The maritime lawyers (and there is a relatively small circle of them) will soon recognise a salvage officer they can professionally trust. And there is a point especially important when it is realised that a learned counsel who is 'on your side' in this case may be on the 'opposite side' in your next salvage case.

I know of one senior salvage officer held in such high esteem by all, that the most eminent lawyers have no hesitation in consulting him officially or unofficially on any doubtful point in any case in which he does not have a direct interest. They regard him as a friend and colleague even if they sometimes have to put him through the hoop when he is on the 'opposite side'.

The worst Salvage Report comes from the salvage officer who starts with the end result and then works backwards step by step to the beginning to show how every decision of his was dead right and according to plan!

Now for a last round-up on my beloved Lloyd's Open Form. There have been – in the last decade – certain criticisms of it; they have come chiefly from foreign salvage operators concerned with the length of time elapsing between a successful case and actually getting their hard-earned cash. Every business person will sympathise and understand that the effects of not knowing 'how much and when?' for incoming funds can affect the current operations and future planning. But there is really such an enormous amount of good in the LOF alongside the offending bit that amendment not abandonment is required.

Arbitration, as already explained, is cheaper and quicker than litigation in any competent court of law – and much more satisfactory than litigation in countries whose courts have no background and tradition of maritime cases. And, if readers would refer back a few pages to the 8 points put forward by Sir William Kennedy, it can be clearly seen that each one of those points requires immense preparation, assessment and detailed documentation. The whole thing is a project requiring painstakingly careful exactitude and I am sure that all salvors would prefer a well considered final decision rather than a quick and subsequently blemished one.

On top of all that, dates have to be arranged far enough into the future so that every essential participant can make sure of his availability to attend proceedings. This is not easy when top experts always have crammed engagement diaries for months ahead.

Obviously, as in any project in any sphere of activity, there may be short cuts to explore, procedures which can be expedited and unnecessary demands eliminated. Lloyd's Open Form has never been set in concrete and is always adjusted to suit up to date demands. The elemental fact, however, that fair decisions cannot be reached without time-consuming detailed information (by both parties) being fed in to the Arbitrator. Lack of sufficient explanatory detail could well adversely affect a claim. It pays to submit too much information rather than too little. Nevertheless, time can and will be saved.

One final note on an associated subject i.e. the growing worldwide pressure on salvors to protect the environment. International salvors going to the assistance of oil tankers and vessels carrying hazardous cargoes (which may not always be known to them at the outset) have great difficulties in bringing their services to a successful conclusion. Harbour-masters resist entry to their ports and governments resist approach to their coastlines.

To keep damaged vessels well offshore in any sort of adverse weather conditions which may happen along is evidently to increase the risk of losing the ship altogether despite all the hard and dangerous work that has been put in already. At least one vessel to my personal knowledge sank in the Western Approaches on the edge of the continental shelf after tugs had held her in an offshore 'wait position' for several days. That was a blow to the No cure – no pay concept.

It is quite clear that, without changes in methods of remuneration, salvors will not be willing to undertake salvage cases of this nature in future. It is no commercial proposition to be left holding an extremely awkward baby at the end of the enterprise and not to get paid for holding it.

Governments walk a desperate and unwelcome tightrope. Because of prevailing westerly weather, any unattended Atlantic casualty (or if the ship sinks, its cargo) is eventually bound to land up on the European coast. The unenviable choice is whether to permit a casualty to make a controlled approach to a chosen port or shoreline which may be far from popular with the local populace or possibly to receive the end result of a damaged ship and cargo into, say, a top holiday resort or profitable industrial conurbation with direst results. It's a case of 'Tails I lose. Heads I lose worse'.

An International Convention on Salvage 1989 has put forward proposals and recommendations. But, as with all international negotiations, it is on the back burner and cooking very slowly.

In the meantime, Lloyd's Open Form has been applauded for amending its terms to guarantee salvors expenses' plus an 'uplift' where their actions succeed in preventing or minimising pollution. That will go a very long way to restoring salvors' morale until inter-governmental conditions can enter into force.

'Reading maketh a full man'; said Francis Bacon, 'conference a ready man and writing an exact man'. Until I started to write about the vast subject of maritime law in a couple of thousand words, I did not realise just how exact I would have to be.

Chapter 14

Salvage, recovery and associated literature

Because this book is in itself small, I have right the way through intended to finish with what is the most useful thing possible – a list of books and periodicals for further reading. I have given much thought to the best method of presentation. Listing under subject-headings is impracticable because many books would require listing under several different headings. Listing under alphabetical order of authors is in no way useful to those who do not know the author's special subjects. Listing of titles in alphabetical order is also impracticable because many of the titles are not self-explanatory.

I suggest therefore that you skim through the books listed below in the order in which I have noted them on my bookshelves. As the titles are frequently not self-explanatory. I have appended a miniature 'review' to each. Then for ease of back reference I will note them in appropriate places in the general index. There is no doubt that this chapter is something of a catalogue but it is nevertheless an interesting and rewarding one. Remember the views of that prolific writer, Anatole France, who said: 'I do not know of any reading more easy, more fascinating, more delightful than a catalogue'.

To save unprofitable and unsatisfying reading I have

suffixed the reviews with (P) when suitable mainly for professional reading, with (G) when suitable mainly for general reading and (PG) when suitable for both classes of reader. Some of the books are out of print but can often be found in second-hand bookshops – especially those who advertise in nautical magazines as specialising in maritime books. Failing that, I have found the British Library a never-failing provider of otherwise unavailable books. Naturally the books have to be asked for in advance and read on the premises since they must not be removed. The British Museum Reading Room is a marvellous facility for students of this (or any other) subject.

Lloyd's Maritime Atlas, Lloyd's of London Press, several editions from 1951 onwards. A useful portable atlas orientated to seaports and shipping places – giving latitudes and longitudes of many places around the coastlines of the world (P).

Ports of the World, Benn. Annual publication containing factual details of repair facilities, crane facilities, pilotage etc. (P).

Brown's Tidal Streams for the Whole of the British Coasts, Brown, Son & Ferguson, Ltd., Glasgow. Hourly charts of tidal stream directions and speeds on small scale maps (P).

Jane's Ocean Technology, Jane's Yearbooks. Up-to-date details of submersibles, habitats, underwater vehicles. Support ships, etc. (PG).

Lloyd's of London, by Raymond Flower and Michael Wynn Jones, David & Charles, 1974. Illustrated history of the great insurance establishment (G).

Brown's Nautical Almanac, Brown, Son & Ferguson, Ltd., Glasgow. The Merchant Navy's annual bible on astronomical ephemeras, tide tables, tidal constants for secondary ports, distance tables, one advertisement for salvage services (P).

The Practice of Ocean Rescue, by Lt-Cdr R.E. Saunders, MBE, RNR, Brown, Son & Ferguson, Ltd., Glasgow, periodically updated. Seaman's guide to deepsea towage (P).

Lloyd's Yearbook, Lloyd's of London Press, annual publication. Contains useful list of Lloyd's Agents at home and abroad. Advertisements from many salvage firms. Examples of Lloyd's policies (P). Formerly called *Lloyd's Calendar*.

Lloyd's List, Lloyd's of London Press, daily newspaper. Topical articles and news. Sailings and arrivals of ships. Casualty reports (P).

Lloyd's Register Book, published annually in two volumes. Essential to any salvage headquarters office and salvage vessel as it contains comprehensive details of the world's seagoing ships of more than 100 tons (P).

Lloyd's Shipping Index, Lloyd's of London Press, daily publication showing movements of some 18,000 ships on overseas voyages. Casualty reports. Essential to any salvage headquarters office (P).

Lloyd's Weekly Casualty Reports, Lloyd's of London Press. Essential only to headquarters offices wishing to keep an easily filed record of all casualties (P).

Admiralty Sailing Directions, Hydrographer of the Navy. A series of volumes covering every part of the world – climatic and other weather conditions, tidal streams, offshore hazards, etc. A set to cover all possible operation areas required in every headquarters and every salvage vessel (P).

Deep Diving, by Robert H. Davis. Published by the great diving firm of Siebe Gorman and periodically updated. Probably the most comprehensive and interesting diving book in print (PG).

Salvage, recovery and associated literature

Marine Salvage, by Joseph N. Gores, David & Charles, 1972. An account of many marine salvage operations by an American engineer (G).

Admiralty Manual of Seamanship, vol. 3, published periodically by HM Stationery Office. Excellent chapters on towage at sea, salvage operations, and handling of ships in heavy weather. Earlier editions had a chapter on wreck dispersal which was omitted after the then Admiralty handed back responsibility for wreck dispersal to the Lighthouse Authorities (P).

Ship Salvage, by Captain G.J. Wheeler, George Philip & Son, 1958. Very understandable but now somewhat outdated descriptions of cases by a first class practical salvage officer (PG).

Epics of Salvage, by David Masters, Cassell, 1953. General book written by a layman for laymen (G).

Vasa, The King's Ship, by Commander Bengt Ohrelius, Cassell, 1962. A book describing the sinking of *Vasa* and her ultimate salvage (G).

The Warship Vasa, by Anders Franzen, Norstadts, Bonniers, Stockholm, 1966. A well illustrated book much concerned with the Marine Archaeology aspects of the warship recovery (G).

The Log of the Neptun Company, Seroco Reklam AB Gothenburg, 1970. Privately printed history of a great salvage company from 1870 to 1970. Including the *Vasa* salvage and many others (P).

Under the Mediterranean, by Honor Frost, Routledge & Kegan Paul, 1963. A well written and well illustrated book on the marine archaeology aspects of ancient wrecks, ships' fittings and harbours (G).

Nautical Magazine, Brown, Son & Ferguson, Ltd., Glasgow. Established 1832. Articles, news items, monthly casualty lists (PG).

Wrecks and Rescue Round the Cornish Coast, by Cyril Noall and Grahame Farr, Bradford Barton Ltd, 1965. Good illustrations (G).

Wonders of Salvage, by David Masters, Eyre & Spottiswoode, 1946. By a layman for laymen (G).

Down to the Ships in the Sea, by Harry Grossett, Hutchinson, 1954. Salvage as personally experienced during the working life of one of the most experienced divers of his time (G).

Subsunk, by Captain W.O. Shelford, RN. FRSA, Harrap, 1960. A first-class account of submarine escapes through the years (G).

The Burning Sea, by Iain Crawford, Cassell, 1959. Interesting semi-fictional account of a wartime salvage tug and its adventures (G).

An Agony of Collisions, by Peter Padfield, Hodder & Stoughton, 1966. Very many collisions incidents recorded, analysed and commented on (PG).

The Sea Surrenders, by Captain W.R. Fell, CMG, CBE, DSO, RN, Cassell, 1960. An autobiographical story of many salvage cases by a distinguished former Admiralty Salvage Officer (PG).

The Man Who Bought a Navy, by Gerald Bowman, Harrap, 1964. The incredible story of a scrap dealer called Ernest Cox who bought all the scuttled German warships in Scappa Flow and successfully learned from scratch the art of ship recovery the hard way (PG).

Underwater Medicine, by Surgeon Rear-Admiral Stanley Miles, CB, MD, MSc, DTM & H and Surgeon Captain D.E. Mackay, RN, MD, DPH, Adlard Coles, 1976. The title is self-explanatory and the book by two of the greatest British experts in diving physiology (P).

The Man in the Helmet, by Desmond Young, Cassell 1963. A story of diving from earliest days to publication date by the son of the greatest British salvage expert of his day (G).

Ordeal by Water, by Peter Keeble, Longmans, 1957. A personal account of wartime salvage mainly in the Eastern Mediterranean by a wartime Admiralty Salvage Officer (PG).

The Wreck of the Memphis, by Captain Edward L. Beach, USN, Jarrolds, 1967. An account of the wreck of a US Battlecruiser in 1916, desperate attempts at salvage, and final abandonment. Written by the son of the then captain of *Memphis* (G).

Up She Rises, by Commander Frank Lipscombe. OBE, RN and John Davies, Hutchinson, 1966. A first-class account of salvage work by the Admiralty Salvage Organisation over the years. Excellent diagrams (PG).

Dictionary of Nautical Words and Terms, by C.W.T. Layton, Brown, Son & Ferguson, Ltd., Glasgow. This title is self-explanatory. Periodically updated (PG).

Encyclopaedia Britannia, published periodically. William Benton. As on so many subjects, many excellent words of wisdom on marine salvage, diving and related subjects. One article leads to another; an education in itself (PG).

Diving Manual, BR 2806, published by Ministry of Defence, Weapons Department, (Naval), and periodically updated. A splendid professional guide to diving practice, regulations, etc., and to all types of diving other than those items which for security reasons cannot be allowed into a book on general sale to the public (P).

US Coastguard in World War 2, by Lieutenant Malcolm F. Willoughby, US Naval Institute, Maryland, 1956. As the title implies a general wartime history of the US Coastguard but in which a few incidents of rescue and salvage occur (G).

Lloyd's of London, by D.E.W. Gibb, Lloyd's of London Press, 1972. A history of Lloyd's by a member who spent a lifetime in the business. Very authoritative (G).

Sound Underwater, by Gregory Haines, David & Charles, 1974. The chapters on Fish Detection, Surveying and Basic Concepts are of educative value to the professional salvor (PG).

West Country Shipwrecks, by John Behenna, David & Charles, 1974. A pictorial record 1866-1973 (PG).

Naval Review, annual publication by USNI, Maryland. Contains an excellent month by month chronology of events at sea. Naturally orientated towards naval affairs but often with interesting entries concerning accidents, towages, submersibles, and other items connected with recovery and salvage (PG).

United States Naval Institute Proceedings, monthly publication by USNI, Maryland. Orientated mainly to naval matters but few issues are without titbits on underwater techniques, submersibles, diving tasks, etc. (PG).

Marine Salvage Operations, by Edward M. Brady, Cornell Maritime Press, Cambridge, periodically updated. A good book by a professional US Salvage Officer. Diving, types of ship-misfortunes, naval architecture, equipment, seamanship tips. Every active and potential salvage officer ought at least to skim through to assure himself that he knows it all! (PG).

Salvage, recovery and associated literature

Law of Civil Salvage, by W.R. Kennedy, regularly revised and updated. Essential for every salvage officer to know the parts which affect him. Essential for the maritime lawyer to know the lot. Certainly the most authoritative book on the subject in Britain – and probably in the world (P).

Hydrographer of the Navy's Publications. Most seamen are well aware of the range and usefulness of Admiralty charts and Admiralty sailing directions. But they are inclined to forget the many other hydrographic publications such as tidal stream atlases and charts, tide tables, sub-surface data, etc. A catalogue is produced every year by the Hydrographer and no professional salvor should fail to acquaint himself with the publications available. The catalogue is held by all Admiralty Chart agents for sale but alternatively a diplomatic word might obtain a free look through on the premises (P).

The Lifeboat, quarterly journal of Royal National Lifeboat Institution, West Quay Road, Poole, Dorset, BH15 1HZ. Available on subscription. Mainly about life-salvage, but some incidents of towage and minor salvage are reported (PG).

Coastguard, official quarterly journal of HM Coastguard, published by Department of Transport. Mostly about life-salvage, collision prevention in Dover Strait and search and rescue co-ordinations (PG).

Safety at Sea International, published monthly by Argus Business Publications Ltd. A journal devoted to promotion of safety in design, construction and operation of shipping. Interesting reports on casualty official inquiries (PG).

Reports of Court, published periodically as cases occur by HMSO in accordance with the Merchant Shipping Act, reports of formal investigations into ship losses (P).

International Shipping and Shipbuilding Directory, annual publication by Benn. Contains separate directory of towage, salvage and offshore services (P).

The Return of the Great Britain, by Richard Goold-Adams, Weidenfeld & Nicolson, 1976. A story of a unique project to rescue Brunel's wonder-ship *Great Britain* from her sunken hulk status in the Falklands and to restore her for show in the UK. One chapter deals with the ingenious lifting of the old ship onto a pontoon and of the subsequent tow home (G).

Shipping Law Titles. Various publications by Lloyd's of London Press. Really for lawyers only and dedicated students (P).

Requiem for a Diver. Warner and Park. Published by Brown, Son & Ferguson, Ltd., Glasgow, 1990. A smallish book which is really a story of what happened to Jackie Warner in his progress from Seaman Boy to Lieutenant-Commander and finally to civilian Chief Inspector of Diving at the Department of Energy. Blunt, uncompromising, statements about the hazards of diving – especially about those for divers on the offshore gas and oil rigs.

Mud, Muscles and Miracles. Captain C. A. Bartholomew USN. Joint publication by US Navy Historical Center and Naval Sea Systems Command, Washington, DC, 1990. A fine selection of interesting salvage cases dealt with by the USN over the years. Very readable, understandable and often thrilling (PG).

US Navy Towing Manual. Published by Naval Sea Systems Command, Washington, DC. Title is self evident. Everything from policy to nuts and bolts. Periodically updated (P).

Tricks of the Trade for Divers. John Malatich and Wayne D. Turner. Published by Cornell Maritime Press, Centreville, Maryland, USA, 1986. Meant for persons who are already qualified divers and have some experience. (P)

Salvage, recovery and associated literature

Ships and Shipwrecks of the Americas. Edited by George F. Bass. Published by Thames and Hudson, 100 Fifth Avenue, New York, 1988. Beautifully written and illustrated (G).

Mary Rose – Excavation and Raising of Henry VIII Flagship, by Margaret Rule. Published by Conway Maritime Press.

How We Found the Mary Rose, by Alexander McKee. Souvenir Press, 1991. A practical account by the leading diver.

Index

F

G

H

Index

In a book such as this, the chapter headings are, in themselves, an index. For example, if you wish to refer back to a refloating case, it is a matter of seconds to go right back to the relevant chapter and flick through the very few pages until the required incident comes to light. However, for persons' names, ship names and certain special items etc., an index is an additional convenience.

Also, as I promised in Chapter 14, a second look at the bibliography — this time in alphabetical order — will probably be extra helpful. However, one special comment is absolutely essential at this juncture. In this bright new world of ours, long-respected publishing firms fold and die. Others go through voluntary or shotgun marriages with other publishers and emerge with strange-sounding names. Seemingly immortal dailies, monthlies and annuals cease publication e.g. that magnificent Liverpool *Journal of Commerce* which was part of our professional lives for so long is unbelievably no more. But you can still call it up in the libraries to see what it said in the past.

Despite difficulties, though, the information I have given should impart sufficient guidance to track down any publication named. Good firms exist which specialise in stocking or finding nautical books which are out of print —

advertisements for such are usually to be found in *Sea Breezes* (202, Cotton Exchange Building, Old Hall Street, Liverpool L3 9LA) and *Nautical Magazine* (4-10 Darnley Street, Glasgow, G41 2SD). Private professional libraries are good if you have access. Public libraries can have some astonishing successes if you give them time. The British Library has never let me down but you need a reader's pass and a little patience. I know that this paragraph is somewhat a repetition of comments at the beginning of Chapter 14 but it is very worthwhile repetition. Guidance is essential even if one wants merely to track down the last issue of a now discontinued publication. An Index which does not give guidance is like a sailor without a knife. And everyone who has ever been to sea knows what a sailor without a knife is like!

In passing, the index-compilation could also be of use to shipowners. Statistically the listings imply that if ships are to have a happy and uneventful life with least chance of featuring in the casualty lists, they should be launched with names beginning with 'I', 'J', 'O', 'Q', 'X' and 'Z'. Alas, just one more case of 'Lies, damned lies and statistics' I reckon.

I have taken five days to put the Index together and, since completion, have tried to 'improve' it. A word here and there to clarify an entry. A double entry now and again under different initials to make sure such is seen. Then, a horrific thought strikes one. If an improvement can be done on certain selected entries, why not to all? This is the sure path to madness. So, as Lady Macbeth so wisely said: 'What's done, 'tis done'. Enough is enough. I hope and think that you will find the result adequate. 'Perfection is not given to man'.

At the very point of seeing proofs for the last time before the book goes to print I have come across a new and specially important publication for lawyers, salvage offices and salvage masters. It is LOF 90 AND THE NEW SALVAGE CONVENTION by the very well known and respected maritime lawyer Gerald Darling Q.C.; it is published by Lloyd's of London Press.

Good sailing!